AN INTRODUCTION TO
INSECT PHYSIOLOGY

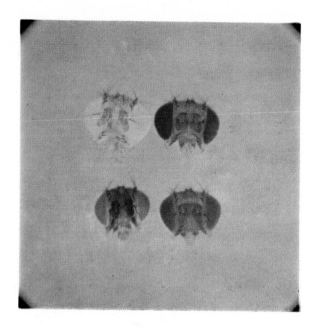

Frontispiece. Eye colours of the fruit fly, *Drosophila melanogaster: top left*, white eye (w); *top right*, sepia (se); *bottom left*, eosin (w^e); *bottom right*, wild type (+). (Photograph courtesy of Mr. P. F. Hulley.)

AN INTRODUCTION TO

INSECT PHYSIOLOGY

by

E. BURSELL

Professor of Zoology, Department of Zoology
University College of Rhodesia, Salisbury
Rhodesia

1970

ACADEMIC PRESS
London and New York

ACADEMIC PRESS INC. (LONDON) LTD.
Berkeley Square House
Berkeley Square
London, W1X 6BA

U.S. Edition published by
ACADEMIC PRESS INC.
111 Fifth Avenue
New York, New York 10003

Library of Congress Catalog Card Number: 73-117144
SBN: 12-146650-7

Printed in Great Britain by
The Whitefriars Press Ltd., London and Tonbridge

PREFACE

The need for another textbook on insect physiology may not seem particularly compelling, as students of this subject are already well served by a number of outstanding texts. These include several excellent review series like *Advances in Insect Physiology* and *Annual Review of Entomology*; a number of good co-operative enterprises among which "Insect Physiology", edited by K. Roeder, and "The Physiology of Insecta", edited by M. Rockstein, may be mentioned; and, bright among these lesser lights, successive editions of Wigglesworth's unrivalled "The Principles of Insect Physiology". It is my experience as a teacher, however, that these books provide too heavy a fare for undergraduates who are not yet fully committed to the study of insect physiology, and it is for such students that the present volume is intended. It attempts to provide a less detailed account of a field which continues to be fruitful for the elucidation of many fundamental aspects of physiological function; and it is written in the hope that a broader approach may better serve to quicken the interest of potential converts.

In order to present a general treatment of insect physiology in a book of manageable size, it has been necessary to assume a certain knowledge of physiology on the part of the reader. Against this assumed background, those aspects of insect physiology which appear to be characteristic of the class have been singled out for detailed discussion. Attempts have been made, wherever possible, to sketch in the experimental background to the information which is presented, by coupling the account of a given topic with appropriate illustrations from the research literature. To these illustrations full reference is provided, giving a point of entry into the literature for readers who may wish to go more deeply into a particular subject; the text itself has been kept free of references, except where it has been thought desirable to draw attention to important work not included among the text figures.

The common names of insects, like cockroach or blowfly, are often used in the text where there seems little to be gained by a more precise designation. This practice does raise the difficulty that there are a number of quite different types of blowfly or cockroach, and what is said of one type does not necessarily apply to another. To avoid possible misinterpretation, an appendix is provided that sets out the common names of insects mentioned in the text, and lists the scientific names of species of that type which have been used as experimental material, together with a page reference to the corresponding item of information.

August, 1970   E. BURSELL

INTRODUCTION

Before setting out to discuss the physiology of insects it may be useful to attempt some definition of what, in the present context, will be meant by the word "physiology", and what by the word "insect", and in this way to delimit the general field of enquiry.

The approach to physiology which I propose to adopt is based on the view that insects, like any other form of animal life, can ultimately be regarded as self-replicating metabolic systems; systems, that is, which possess the catalysts and cofactors necessary to promote a particular pattern of transformation of energy and material, to sustain a particular type of metabolism. Usually such a pattern involves the breakdown of complex organic molecules, with capture of a part of the energy so released in useful form; and the synthesis, usually from simpler molecules, of components of the metabolic system itself. In order that such a system shall continue to exist a number of requirements must be fulfilled. There must, for instance, be a continual supply of complex molecules to serve as a source of energy and as raw material for synthetic purposes; oxygen must be supplied to meet the needs of oxidative degradations; the catalytic machinery must be maintained in an environment suitable for its activity, necessitating the removal of toxic end products and the regulation of water content and ionic composition. The term somatic physiology may be used to denote the processes by which these different requirements are met. It would include, for example, the processes of nutrition, digestion, respiratory exchange, excretion and osmoregulation, studied largely at the level of organ systems, such as the alimentary canal, the tracheal system and the excretory system. In addition there are activities of the organism as a whole which tend to the fulfilment of metabolic requirements, and here one would be concerned with a study of the behaviour of the organism in relation to its environment, as mediated by the neuromuscular system. Finally, there are those aspects of physiology which relate particularly to the self-replicating nature of the system, and these may be considered under the separate heading of reproduction and development. Genetical aspects of replication have, somewhat arbitrarily, been deemed to lie outside the scope of the present book.

In what follows, the word "physiology" will thus be used to denote the sum total of processes tending to the maintenance and replication of a metabolic system. And this book aims to deal with that kind of metabolic system which is called an insect. There are, however, over a million species of insect, and between them they show an amazing diversity of adaptation to widely different modes of life, many of them involving profound specializations at the physiological level. One could therefore envisage the existence of a large number of types of insect physiology, and in a general treatment some sort of simplification would be essential. The question arises as to the basis on which such simplification could reasonably be made. One way would be to limit consideration to what one might hopefully regard as a "typical" insect. But apart from the difficulty of deciding what kind of animal this would be, one is faced with the fact that our knowledge of insect physiology is somewhat fragmentary. Certain aspects of physiology are well documented for certain insects, others for others, but in no single species are we in a position to build up anything like a complete picture. In view of this, a better approach might be to limit consideration as far as possible to those aspects of described physiology which could be thought to be typical of insects generally, rather than just of a particular species. Here an element of subjective judgement is involved, but since simplification must inevitably mean selection of some kind, this cannot be avoided. What can be done, however, is to elaborate a little on the sort of criteria which could be used as a basis for selection; to consider, in other words, what are the characteristics of insects generally which would be most likely to affect their physiology. Here one would perhaps list, in the first place their terrestrialness, with all this implies in terms of desiccation, insolation, thermal fluctuations and so on. There is, secondly, their smallness; the largest insects are barely as big as the smallest terrestrial vertebrates, while the smallest insects are little bigger than many protozoans. This means that the surfaces available for exchanges with the environment are large in relation to the volume which serves as the source or sink of the exchange. This surface/volume relationship has wide implications in relation to such aspects as water balance, heat balance and respiratory exchange. The capacity for flight is another characteristic of insects which could be expected to have far-reaching physiological implications, with particular reference to neuromuscular physiology and to the mobilization of metabolic reserves. Fourthly, and perhaps most importantly, there is the fact that insects are exoskeletal; that their cuticle must play a dual role, as a skeleton and as a protective layer. This is a feature of very great consequence, particularly in connection with growth and development, and with aspects of metabolism associated with the deposition of the cuticle.

It would be possible to extend this list very considerably, but enough has perhaps been said to indicate the sorts of features which seem likely particularly to affect insect physiology, and which have therefore hopefully been used as a guide in deciding what is relevant to a general discussion of insect physiology, as against what may be regarded as a peculiar characteristic of this or that particular species of insect.

ACKNOWLEDGEMENTS

I am greatly indebted to the following friends who have been so kind as to read and comment most helpfully on different sections: Professor E. B. Edney, Professor L. H. Finlayson, Dr. J. P. Loveridge, Dr. M. P. Osborne, Dr. R. J. Phelps and Mr. D. J. W. Rose. My special thanks are due to Dr. C. B. Cottrell who devoted a great deal of time to the improvement of Section III. Readers will have cause to be thankful to my wife, Mercia, who has made a valiant attempt to improve my English; to her this book is dedicated.

For permission to reproduce material from a number of publications I wish to express my gratitude to all the publishers concerned.

CONTENTS

SECTION I: SOMATIC PHYSIOLOGY

SECTION II: NEUROMUSCULAR PHYSIOLOGY

SECTION III: THE PHYSIOLOGY OF REPRODUCTION AND DEVELOPMENT

SECTION IV: ASPECTS OF PHYSIOLOGICAL ECOLOGY

CONTENTS

SECTION I

Somatic Physiology

CHAPTER 1

METABOLISM

Metabolism may be considered as being of two main kinds; one involving the breakdown of complex organic molecules and the trapping of part of the energy which they contain in high energy phosphate linkage; the other involving a synthesis of complex organic molecules from simpler ones, to replace or renew components of the metabolic system. The two kinds have many enzymes and pathways in common, and are not to be considered as distinct entities, but they do differ in general direction and may conveniently be separated for purposes of discussion.

1. The Breakdown of Complex Organic Molecules

Three main classes of compounds, carbohydrates, proteins and fats are used as sources of energy in insects, as in most other animals. A general outline of the main pathways, showing the interrelationships between the three classes is illustrated in Fig. 1.1, which will serve as a framework for the more detailed discussion which follows.

The focal point of degradation metabolism is the Krebs cycle. This is a complicated system of enzymes, co-factors and substrates which effects the complete oxidation of 2-carbon fragments to carbon dioxide and water. The operation is achieved through a series of decarboxylations and dehydrogenations. The carbon dioxide is released into solution, while the hydrogen which is freed from substrate combination is handed down through a series of hydrogen or electron carriers, of which nicotine adenine dinucleotide (NAD), the flavoproteins and the cytochromes are familiar examples. The electrons are eventually passed on to molecular oxygen with the formation of water, as shown at the bottom of Fig. 1.1. This transport of hydrogen is coupled with a process of phosphorylation, so that a substantial proportion of the energy released in the process of dehydrogenation is captured in high energy phosphate linkage ($\sim$P) leading to the formation of adenosine triphosphate (ATP) from adenosine diphosphate (ADP). The high energy of the terminal phosphate radicle represents what may be called usable energy, and $\sim$P constitutes the common

3

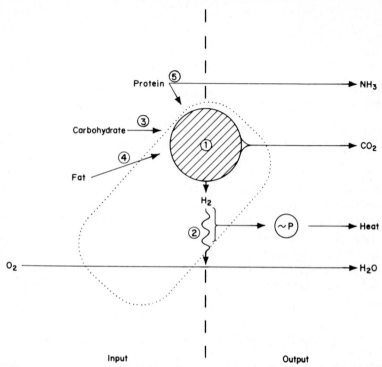

Fig. 1.1. An outline of metabolism. 1, Krebs cycle; 2, hydrogen transport; 3, glycolysis; 4, β-oxidation; 5, deamination; the dotted line delimits mitochondrial events; for further explanation see text.

coin of energy expenditure, being used in virtually all of the energy demanding processes requisite to the maintenance of the metabolic system as a whole, to the maintenance, that is, of the living state. Thus ATP is required for the contraction of muscle, for a variety of chemical syntheses, for the performance of osmotic work, for the production of light in luminescent organs, and so on. The precise way in which the energy of the phosphate bond is transformed into various other forms of energy is still largely unknown, but for present purposes it is enough to accept that the production of ATP, which occurs mainly as a result of oxidative phosphorylation, is an indispensable condition of life.

The input to this central part of the metabolic system is oxygen and foodstuff, as shown on the left side of Fig. 1.1. Carbohydrates are degraded to a form suitable for entry into the Krebs cycle in the process known as glycolysis. This constitutes a fragmentation of the 6-carbon molecule, which arises by hydrolysis of polysaccharides like glycogen, and involves initial phosphory-lations, which serve to bring the molecule into a reactive state, and so facilitate the splitting into two 3-carbon fragments. Each step in the process is catalysed by its own enzyme, as is the case in nearly all of the metabolic transformations under discussion. After further modification the 3-carbon fragment undergoes a process of oxidative decarboxylation in combination with coenzyme A (CoA), to yield the 2-carbon fragment, still linked to the enzyme as acetyl CoA; and it is in this form that it enters the Krebs cycle, to become involved in the process of oxidative phosphorylation. Except for the final decarboxylation, the breakdown of glucose to its 2-carbon derivative takes place in the absence of oxygen, i.e. it is anaerobic.

The fats undergo preliminary hydrolysis to yield their constituent parts, glycerol and fatty acids. Glycerol is phosphorylated to join the glycolytic pathway, while the fatty acids are subjected to a process known as β-oxidation. This involves the splitting off of successive 2-carbon fragments, which takes place in combination with CoA to yield successive molecules of acetyl CoA, of which the 2-carbon fragment enters the Krebs cycle.

The proteins are hydrolysed to their constituent amino acids. Deamination of three of the most common of such amino acids—glutamic acid, aspartic acid and alanine—yield three keto-analogues which are constituents of the metabolic pathways described above, namely α-ketoglutaric acid and oxaloacetic acid (Krebs cycle substrates), and pyruvic acid (glycolytic end-product), thus providing three points of entry to the system. Many other amino acids can be converted to one or other of the three, and in this way the bulk of protein foodstuff can be broken down to yield energy in phosphate linkage. The main difficulty is that the nitrogen which these substances contain has no place in the pathways described, which means that ammonia arises as a toxic end-product as indicated in Fig. 1.1.

There is a close relation between the different parts of the metabolic system and cellular architecture. Thus the process of oxidative phosphorylation is associated with the mitochondrion of the cell, and so is the oxidation of fatty acids. In Fig. 1.1, this association has been illustrated by enclosing relevant parts of the system within a dotted line representing the outer wall of the mitochondrion. The enzymes of the glycolytic pathway, on the other hand, are not associated with a structural framework, but occur free in solution in the cytoplasm, as do many of the transaminases involved in the interconversion and the deamination of amino acids, which precede their entry to the Krebs cycle.

Figure 1.1 illustrates that the continued production of high energy phosphate is dependent on the sustained input of one or more of the classes of foodstuff and of oxygen; and that the output from the system includes toxic products like ammonia and carbon dioxide, which must be detoxicated or removed, and heat, which must be dissipated. Water should strictly speaking figure as an input as well as an output, since many of the reactions concerned in the oxidation of foodstuffs involve a preliminary hydrolysis, that is, an addition of water to the molecule. However, the amount of water produced in the final stages of oxidation greatly exceeds the input, so that in the overall reaction there is a net output, as represented in the figure. In view of the dependence of all metabolic systems on an aqueous medium as a basis for their function, this appearance of water as an end-product of metabolism may usually be regarded as an advantage. The quantity of "metabolic" water so produced differs according to the type of material oxidized, and the possible implications of such differences will be considered later in relation to the question of water balance.

With this general summary of metabolism as a basis, we may proceed to consider firstly the nature of insect metabolism, and secondly the various physiological processes like nutrition, respiration, excretion, osmoregulation etc., which serve to provide an appropriate input to the system, to deal with the output and in other ways to ensure the proper functioning of the metabolic machinery.

2. The Breakdown of Organic Molecules in Insects

Because of the small size of insects, and the correspondingly small quantity of material available for analysis, the study of insect biochemistry has not yet advanced to the level of comparable vertebrate studies. With the development during the present century, however, of a variety of microtechniques, notably those based on paper and thin-layer chromatography, it has become possible to isolate and assay extremely small amounts of different metabolites, and a picture of insect biochemistry is beginning to emerge as a result of these technical advances.

a. The Insect Mitochondrion

During recent years a great deal of work has been done on the insect mitochondrion, particularly with the flight muscle mitochondrion usually referred to as the sarcosome. These mitochondria constitute particularly favourable experimental material because of their large size, and their occurrence in closely packed rows between the intracellular myofibrils of the large muscle cells (see Fig. 1.2). Their isolation, in amounts adequate for experimental work, is easily and rapidly accomplished by grinding the thoraces of suitable insects with pestle and mortar, filtering off the coarse particulate material comprising cuticular and tracheal fragments mixed with partially dissolved myofibrils, through layers of muslin cloth, and spinning down the large sarcosomes at quite low centrifugal speeds. After washing of the sarcosome pellet and re-suspension in a suitable medium the properties of the mitochondrion can be studied; the rate of oxidation of various substrates can be determined on the basis of oxygen consumption, either by standard manometric methods, or with the more recently developed oxygen electrode; the oxidation and reduction of components of the hydrogen transport system can be studied by spectrophotometric

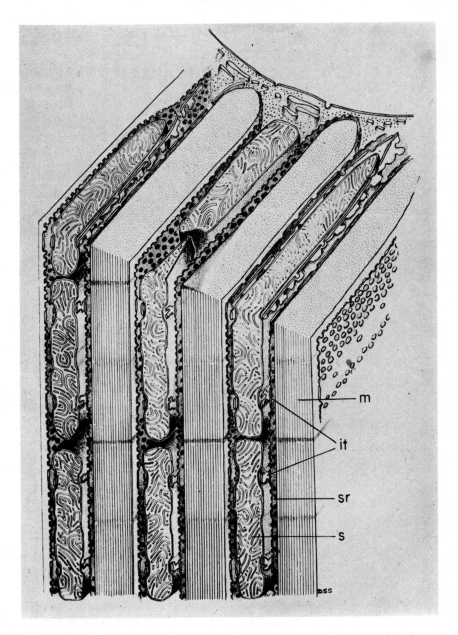

Fig. 1.2. Diagrammatic reconstruction of fine structure in the flight muscle of the dragon fly. it, intermediary tubule; m, myofibril; s, sarcosome; sr, sarcoplasmic reticulum (Pringle, 1965 after Smith).

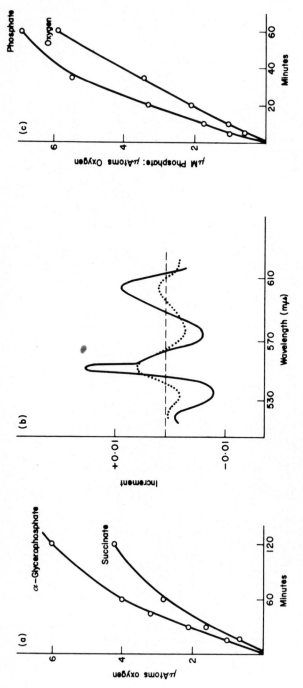

Fig. 1.3. Three different approaches to the study of mitochondrial activity. (a) The manometrically determined oxygen consumption of teased muscle preparation of the locust, showing the high rate of oxidation when αglycerophosphate is provided as a substrate (data from Zebe et al., 1959). (b) The state of oxidation of mitochondrial constituents as shown by the split-beam spectrophotometer; the dashed curve represents the increment of absorption over a range of wavelengths between aerobic sarcosomes of the housefly in the absence of substrate (hydrogen carriers fully oxidized) and in the presence of α-glycerophosphate. The solid curve shows the difference spectrum in the absence of oxygen (hydrogen carriers fully reduced). The absorption peak of reduced cytochrome c (at 550 mμ) and of cytochrome a (at 605 mμ) are clearly seen under the steady state conditions of oxidation, showing that these pigments play a part in hydrogen transport during oxidation of α-glycerophosphate (Chance and Saktor, 1958). (c) The esterification of phosphate and the uptake of oxygen by sarcosomes of the blowfly during oxidation of α-ketoglutarate. The P/O ratio (see text) averages about 1.5 for the experiment (Lewis and Slater, 1954).

means; and the ability to generate high energy phosphate bonds can be followed by determining the anhydride formation of inorganic phosphate with ADP to produce ATP (see Fig. 1.3).

In this context it should be noted that the mitochondrion is an extremely complex and delicate organelle, and one which is very liable to degenerative changes. During isolation careful precautions must be taken to guard against deleterious influences such as osmotic shock, excess calcium, accumulation of inhibitory factors etc. In fact, the properties of the mitochondrion depend to a rather uncomfortable extent on the method used for its isolation; even with all recommended precautions taken, one may hesitate to equate *in vitro* with *in vivo* performance, and experimental results have usually to be accepted with reservation.

All the enzymes of the Krebs cycle have by now been shown to occur in the mitochondria of different insects, and the ability of the insect mitochondrion to effect the complete oxidation of pyruvate has been amply confirmed. A number of other substrates have been tried, and α-glycerophosphate has been found to be a particularly good one, capable of being oxidized faster than pyruvate, and than members of the Krebs cycle itself, such as succinate (see Fig. 1.3(a)). This difference will be discussed further when the glycolytic pathway comes up for consideration.

The electron transport system has also been extensively studied particularly since the development of the split-beam spectrophotometer, which enables measurements to be made of the state of oxidation of hydrogen carriers in the intact mitochondrion (see Fig. 1.3(b)). The carrier system appears to conform to the basic pattern, as illustrated in Fig. 1.4. Hydrogen is passed on from substrate to flavoprotein, and from there through a series of cytochromes to oxygen. Three different flavoproteins have been shown to be involved in the transfer of electrons from different substrates to cytochrome *b*, their relative activities varying from tissue to tissue. In sarcosomes the α-glycerophosphate pathway appears to be particularly active, in accord with the results reported above, and another carrier, coenzyme *Q* appears to be interposed between the flavoprotein and cytochrome *b*.

It is generally agreed that during the transfer of electrons from NADH to oxygen by a carrier system of this sort, the reduction of one atom of oxygen is accompanied by the generation of three molecules of ATP from ADP and inorganic phosphate. Where the hydrogen is handed on directly from substrate to flavoprotein, as for succinate or α-glycerophosphate, the first phosphorylation, which occurs at the level of NADH oxidation, is lost, giving a theoretical maximum of two molecules of ATP per atom of oxygen. These phosphate/oxygen ratios, usually denoted as P/O, have been determined for a range of insect mitochondria, and it has proved possible during recent years to attain to values which approximate quite closely to the theoretical maximum (see Fig.

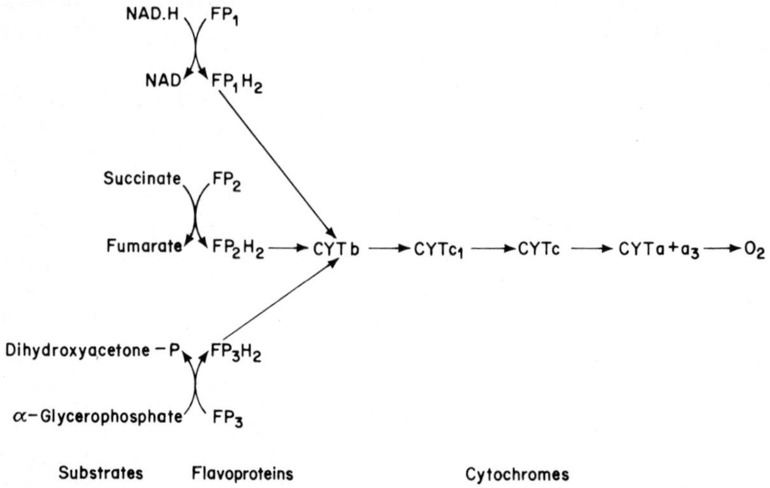

NAD.H   FP$_1$

NAD   FP$_1$H$_2$

Succinate   FP$_2$

Fumarate   FP$_2$H$_2 \longrightarrow$ CYT b $\longrightarrow$ CYTc$_1$ $\longrightarrow$ CYTc $\longrightarrow$ CYTa+a$_3$ $\longrightarrow$ O$_2$

Dihydroxyacetone − P   FP$_3$H$_2$

α−Glycerophosphate   FP$_3$

Substrates   Flavoproteins   Cytochromes

Fig. 1.4. Pathways of hydrogen and electron transport in the insect mitochondrion (from Gilmour, 1961).

1.3(c)), although results are rather variable. This may be a reflection of the difficulty of maintaining full functional integrity of mitochondria during the period required for isolation.

The problem of respiratory control has been the subject of a number of investigations with insect material, being particularly sharply posed with the mitochondrion of flight musculature; but the mechanism which is involved in the transition from the resting to the active state, associated sometimes with an increase in metabolic rate of more than 100-fold, is still not fully understood. With mammalian preparations the concentration of ADP has been shown to be of major importance in the regulation of oxidative phosphorylation. When work is done, as during muscular activity, ATP is converted to ADP, and the resulting increase in ADP concentration stimulates mitochondrial activity, and hence the reconstitution of ATP. With insect sarcosomes, too, ADP appears to exercise a measure of control, as shown in Fig. 1.5; during recent years respiratory control ratios (the ratio of rates of oxidation in the presence and absence of ADP) in the region of 30 have been demonstrated, but this still falls somewhat short of the degree of control shown by the mitochondrion *in vivo* during the transition from rest to flight. The remaining discrepancy could be accounted for in various ways, but none of the explanations so far advanced have been convincingly supported by experimental evidence. One possibility would be that the availability of glycolytic substrate for entry into the Krebs cycle might constitute a limiting factor; another that the level of concentration of the Krebs cycle substrates themselves might constitute a limitation, and that their concentration might be

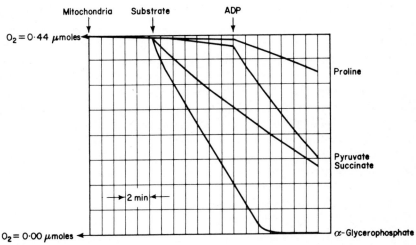

Fig. 1.5. Polarographic oxygen electrode records showing the decrease in oxygen concentration during incubation of blowfly mitochondria with different substrates. 0.1 mg of mitochondrial protein was added to 2.0 ml of a sucrose medium containing 1 mM Mg^{++} at the first arrow; oxygen uptake is negligible in the absence of substrate. At the second arrow the listed substrates were added at a concentration of 10 mM; α-glycerophosphate and succinate are oxidized at high rates, but oxygen consumption in the presence of pyruvate and proline remain low until ADP is added (at the third arrow); both show high respiratory control indices, while succinate and α-glycerophosphate oxidation is unaffected by addition of ADP (from unpublished records).

boosted during early phases of flight by deamination of appropriate amino acids whose keto-analogues could enter the cycle, thus making more oxaloacetate available for condensation with acetyl CoA, and hence allowing a higher rate of input to the Krebs cycle. It is possible that a combination of regulative mechanisms, of which the three here outlined may be examples, co-operate to achieve the high level of respiratory control which characterizes the insect sarcosome; on the other hand, it may be that failure to achieve the requisite level of control with ADP may be attributable to a process of deterioration during sarcosome isolation.

It is only during recent years that information has begun to become available on the oxidation of fatty acids, which constitute one of the main food reserves of many species of insect. Early work had shown that acetate could be oxidized by isolated mitochondria at reasonable rates, but it had not been possible to demonstrate substantial oxidation of higher fatty acids. Subsequently, it was discovered that carnitine esters of long-chain fatty acids were rapidly oxidized by insect mitochondria, and that the addition of carnitine to mitochondrial suspensions greatly enhanced the rate of oxidation of higher fatty acids, pointing to the existence in the mitochondrion of a system capable of synthesizing carnitine esters. The anomalous situation, that insect mitochondria appeared unable to oxidize one of the most important of their natural substrates, has thus been resolved, but further work needs to be done to establish the precise role of carnitine and to extend the observations to a range of insect species.

Before leaving the subject of the insect mitochondrion, it may be well to sound a note of caution concerning the results obtained with isolated preparations. One of the most striking features of the history of investigations in this field is the apparent change in the properties of the sarcosome which has gone hand in hand with refinements of technique, particularly in relation to extraction media and procedure. Early sarcosomal preparations showed very poor oxidative capacity, and there appeared to be little discrimination by the mitochondrion between different substrates. There followed a period during which an apparent distinction could be made between the high rates of oxidation of α-glycerophosphate on the one hand and most other substrates on the other. With later preparations it has been shown that substrates like pyruvate, proline, NADH and succinate may be oxidized at rates equal to or exceeding those of α-glycerophosphate. Similar trends can be seen in relation to other mitochondrial properties; early investigators were unable to demonstrate substantial control by ADP, but at a later stage respiratory control indices were obtained that were little inferior to those which had been found to characterize mammalian preparations, while at the present day it is possible to isolate sarcosomes which show levels of control far in excess of the mammalian mitochondrion. It may well be, however, that with further refinements of technique, based on a better understanding of the phenomena under investigation, we may have yet again to revise our views on *in vivo* mitochondrial function.

b. Glycolysis

Glycolysis in insects, as in other animals, is mediated by enzymes associated with the soluble fraction of tissue homogenates. If the flight musculature of an insect, for instance, is homogenized in water and then subjected to high speed centrifugation, all particulate material can be spun down, not only gross fragments, nuclei and mitochondria, but also a finely particulate "microsomal" fraction, representing remains of the intracellular endoplasmic reticulum. The clear supernant then contains only soluble material, including certain enzymes and co-factors and most of the substrates. It is in this solution that enzymes mediating the breakdown of glycogen to 3-carbon fragments can be found. To demonstrate their presence, the soluble extract is usually dialysed in order to remove endogenous substrates. A particular substrate, for instance glucose-6-phosphate can then be added to the extract, and its transformation to the reaction product glucose-1-phosphate demonstrated, indicating the presence of the corresponding enzyme phosphoglucomutase. Since dialysis of the soluble fraction removes not only substrates but also other small molecules in solution, such as NAD, ATP, inorganic phosphate etc., it is necessary to add appropriate co-factors and ions in order to demonstrate the occurrence of those reactions that require them.

In this way all enzymes of the glycolytic pathway have been shown to be active in extracts from a variety of tissues and from a number of different species of insect. These investigations have also demonstrated one of the most striking peculiarities of insect metabolism, the importance, namely, of a special carbohydrate, trehalose, which occurs as a side-branch of the normal metabolic pathway; this substance appears to play a far more important part in the metabolism of insects than it does in other animals which have been investigated.

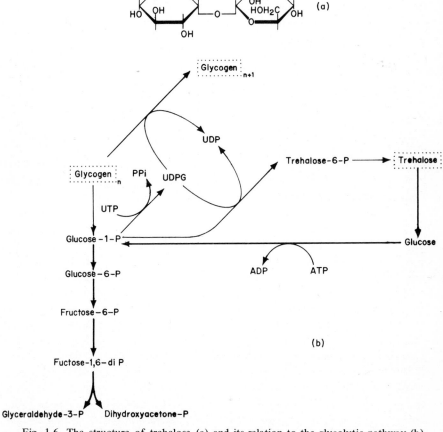

Fig. 1.6. The structure of trehalose (a) and its relation to the glycolytic pathway (b). UTP, uridine triphosphate; UDP, uridine diphosphate; UDPG, uridine diphosphoglucose; n, number of glycosyl units; PP_i, inorganic phosphate.

The structure of trehalose, and its relation to the general glycolytic pathway is illustrated in Fig. 1.6. Trehalose is a disaccharide composed of two glucose units joined through a 1,1 α-linkage. It can be hydrolysed to its constituent glucose by the enzyme trehalase, which has been shown to be active in many insect tissues, particularly high activities being recorded in the alimentary canal. Trehalase is also capable of catalysing the synthesis of trehalose, but this pathway is probably not of physiological significance. As shown in the figure, the normal precursor for synthesis is glucose phosphate and uridine diphospho-glucose (UDPG), which is a complex of glucose with uridine diphosphate, a nucleotide analogous to ADP, but with uridine instead of adenine as the organic base. This complex is a normal intermediary in the synthesis of glycogen, constituting the mechanism by which glucosyl residues are added to pre-existing glycogen chains. This nucleotide complex may, under the influence of the enzyme trehalose-6-P-synthetase, effect the coupling of a glucosyl unit to glucose-6-phosphate, instead of to glycogen, to produce trehalose-6-phosphate, and this, in turn, can be dephosphorylated by trehalose-6-phosphatase to yield trehalose itself.

Trehalose makes up a considerable proportion of the carbohydrate reserve in insects, and occurs in particularly high concentration in the blood, or haemolymph, of resting insects, where values between 0.5 and 5.0 g/100 ml of blood are commonly encountered (see Chapter 3). During flight there is a sharp fall in trehalose concentration, and this, together with a great deal of other evidence, suggests that trehalose constitutes a readily available substrate for metabolism generally and for flight metabolism in particular, as indicated by heavy arrows in Fig. 1.6. The occurrence of this subsidiary reserve in the haemolymph stands in marked contrast to the situation in vertebrates, where blood sugar is maintained at a low level of concentration (usually less than 0.1 g/100 ml). Such a substantial haemolymph reserve would be of particular significance in relation to the heavy demand for substrate which would characterize active flight musculature, especially in view of the relatively inefficient open type of circulatory system in insects (Chapter 3), for it would ensure that a steep concentration gradient would be available to promote the rapid diffusion of substrate across the sarcolemma to reach the oxidative machinery. A simple requirement for rapid diffusion would, however, be better met by a high concentration of glucose, since the glucose molecule is smaller than that of trehalose, and therefore capable of faster diffusion. Considerations of this kind could not, therefore, account for the introduction of trehalose as a respiratory substrate. It is possible that the occurrence of trehalose as the main blood sugar is related not so much to the utilization of carbohydrate as to its uptake from the gut. The presence of high concentrations of glucose would militate against the absorption of this food material, which is a major element of the diet in many insects. Further discussion of this possibility will be deferred till the problem of digestion is considered (Chapter 2).

The occurrence of trehalose as a major element of carbohydrate metabolism is not the only peculiarity of the glycolytic system in insects. What may be considered as a second side-branch of the main reaction sequence is particularly well developed in this group, namely the reduction of dihydroxyacetone phosphate to α-glycerophosphate under the influence of the enzyme α-glycerophosphate dehydrogenase. This enzyme is particularly active in flight musculature, where its occurrence appears to be linked with an apparent defect of the mitochondrial system to which reference has already been made; the inability, that is, of sarcosomes to effect the rapid oxidation of NADH.

One of the steps in the glycolytic transformation of triosephosphate to pyruvate involves a dehydrogenation with NAD acting as hydrogen acceptor. Since NAD is present only in catalytic amounts, the reduced form must be oxidized as rapidly as it is formed if the process of glycolysis is to continue. With the high activity of α-glycerophosphate dehydrogenase in sarcoplasm this re-oxidation can readily be achieved, but in itself this would confer little advantage since, as can be seen from Fig. 1.7, the net yield from the reaction would be zero; the two molecules of ATP which are required for the early phosphorylations are just balanced by the yield of ATP involved in the transformation of triosephosphate to pyruvate, and the NADH formed at one point of the reaction sequence would have to be oxidized at the expense of a reduction of dihydroxyacetone phosphate to α-glycerophosphate. It is only

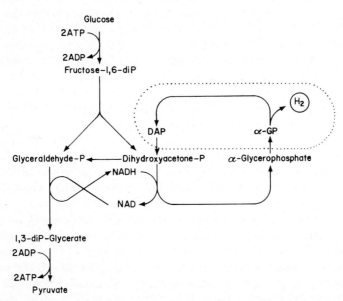

Fig. 1.7. The α-glycerophosphate shuttle system. DAP, dihydroxyacetone phosphate; α-GP, α-glycerophosphate. The dotted line represents the mitochondrion (for further explanation see text).

when this scheme is seen in relation to the capacity of sarcosomes for rapid oxidation of α-glycerophosphate that its significance becomes clear. The α-glycerophosphate formed in the sarcoplasm penetrates into the mitochondrion where it is oxidized to dihydroxyacetone phosphate under the influence of the mitochondrial enzyme (see Fig. 1.3(a) above). The dihydroxyacetone phosphate diffuses back into the sarcoplasm, where it becomes available for the oxidation of another molecule of NADH. What is effectively happening is that α-glycerophosphate serves to carry hydrogen into the mitochondrion, where it can be harnessed to the process of oxidative phosphorylation. Since this transport is in the nature of a shuttle service, it can operate on the basis of relatively small quantities of carrier, so that the major proportion of triosephosphate can be directed to the formation of pyruvate, and the over all reaction sequence becomes

$$\text{glucose} \rightarrow 2 \text{ pyruvate} + 2\text{ATP}$$

with normal glycolytic yields of energy, instead of as before

$$\text{glucose} \rightarrow \text{pyruvate} + \alpha\text{-glycerophosphate}$$

with zero yield of energy.

It may be permissible to speculate that, in the operation of this system, we see another adaptation to the requirement for a rapid release of energy necessitated by the activity of flight. The rate at which reduced coenzyme can penetrate the mitochondrion and become oxidized, while adequate for normal energy requirements, might not be capable of supporting the 100-fold increase in metabolic rate which characterizes flight; under these circumstances there would be need for a special mechanism to ensure a rapid transfer of hydrogen to the oxidative machinery; the simpler alternative, which would be to increase the permeability of the mitochondrion to NAD, might be impracticable in view of the necessity to maintain high concentrations of NAD inside the mitochondrion for intramitochondrial hydrogen transport.

c. The Oxidation of Amino Acids

The protein amino acids occupy a special place in degradation metabolism in so far as they contain between 10% and 34% of nitrogen. This nitrogen must be removed before they can enter the paths of oxidative metabolism and their energy trapped in usable form. The utilization of amino acids as a source of energy in insects is of particular interest in view of the high concentration of amino acids which characterizes insect haemolymph. There is considerable variation between and within species, but it is generally of the order of 1.5 g/100 ml as compared with values for vertebrate blood of about 0.03 g/100 ml (see Chapter 3).

A considerable amount of careful work has been done on quantitative aspects of amino acid composition in insect haemolymph, but the results do not lend themselves to any sort of convincing generalization. With few exceptions, any one of the normal protein amino acids may occur in quite high concentration in one insect or another, and details of the amino acid pattern may vary widely from species to species and between different stages in the life history of a single species. What may be of significance in relation to an aspect of respiratory metabolism to which reference has already been made is that proline (strictly speaking an imino rather than an amino acid) usually figures as a dominant component, often with glutamic acid as its nearest rival, and between them reaching levels of 1.5 g/100 ml (cf. trehalose at 0.5-5.0 g/100 ml). Since an enzyme system which catalyses the oxidation of proline to glutamic acid has been shown to be active in insect mitochondria, and since the deamination of glutamic acid to α-ketoglutaric acid, a member of the Krebs cycle, is readily accomplished by many insect preparations both soluble and particulate, it may be that these two substances, proline and glutamic acid, should be regarded as Krebs cycle "primers", capable of being drawn into the Krebs cycle at the onset of flight and so augment the capacity of this enzyme/substrate complex for pyruvate oxidation. In the insects whose flight metabolism has been investigated it has been shown that the concentration of haemolymph proline declines sharply during early phases of flight, in accord with such an interpretation.

The main prerequisite of amino acid utilization in general is the removal of nitrogen, which can be accomplished by a process of oxidative deamination under the influence of certain flavoprotein oxidases, specific for each of the amino acids. However, the available evidence suggests that this may not be an important pathway in insects. Amino acids are optically active substances which may occur in two isomeric forms, the dextro- and the laevo-rotatory (D- and L-forms), and to these there corresponds the appropriate D- and L-amino oxidases. Naturally occurring amino acids are almost exclusively of the L-form, yet L-amino acids are deaminated slowly, if at all, by insect tissue extracts, while D-amino oxidases are relatively active. Whatever the significance of the occurrence of D-amino oxidation, it is clear that oxidative deamination of endogenous amino acids does not contribute materially to their metabolism. The only exception is glutamic acid which is oxidatively deaminated by a highly active glutamic dehydrogenase, present in both soluble and particulate fractions, and with NAD as the hydrogen acceptor

$$\text{NAD} \quad\overset{\displaystyle\text{glutamate} + H_2O}{\underset{\displaystyle\alpha\text{-ketoglutarate} + NH_3}{\bigtimes}}\quad \text{NADH}$$

This system occupies a central position in the amino acid metabolism of insects, by virtue of the existence in most insect tissues of a variety of transaminases—or amino transferases—catalysing the transfer of amino groups from amino- to keto-acids. When α-ketoglutaric acid acts as the amino-acceptor, and the glutamic acid so formed is oxidatively deaminated, the over-all effect is to provide for the oxidative deamination of the original donor molecule, e.g.

$$\text{alanine} + \alpha\text{-ketoglutarate} \xrightarrow{\text{transaminase}} \text{pyruvate} + \text{glutamate}$$

$$\text{glutamate} + \text{NAD} + H_2O \xrightarrow{\text{dehydrogenase}} \alpha\text{-ketoglutarate} + \text{NADH} + NH_3$$

$$\overline{\text{alanine} + \text{NAD} + H_2O \longrightarrow \text{pyruvate} + \text{NADH} + NH_3}$$

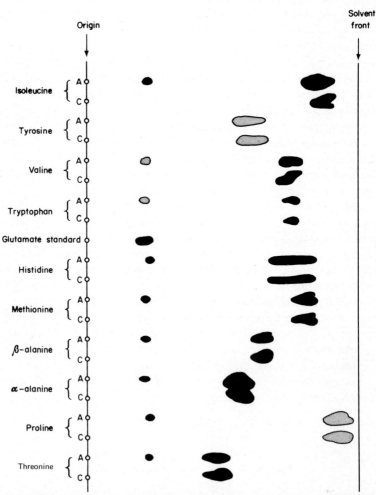

Fig. 1.8. Paper chromatogram showing the transamination of a number of amino acids with α-ketoglutaric acid, leading to the formation of glutamate, by cockroach homogenates. A, active test; C, control containing boiled homogenate (McAllan and Chefurka, 1961).

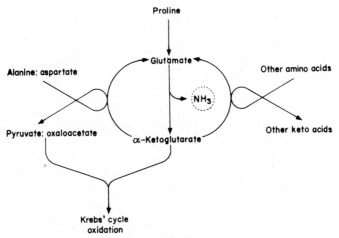

Fig. 1.9. Summary of amino acid degradation, to illustrate the central importance of glutamic dehydrogenase. For further explanation see text.

Transaminases active with α-ketoglutarate as an amino acceptor have been demonstrated in a number of different insects and operative with a number of different donor amino acids (see Fig. 1.8). The enzymes which catalyse the transfer of amino groups from aspartate and from alanine, whose keto-analogues, oxaloacetate and pyruvate, occupy positions close to the centre of degradation metabolism, are usually considerably more active than those involving other amino acids. But whatever the particular amino acid, it seems likely that this sort of coupling of transamination with oxidative deamination of glutamic acid, as summarized in Fig. 1.9, constitutes the main link between the amino acid pool and the energy releasing process of oxidative phosphorylation.

3. The Intermediary Metabolism of Insects

So far discussion has been limited to the breakdown of different kinds of organic compounds with capture of a proportion of the energy released during breakdown. A lot of metabolic activity, however, is directed not to the breakdown of compounds but to their synthesis—to the elaboration of special kinds of material disposed within the cell or secreted from it—and required for a variety of special purposes. Metabolic transformations of this kind may conveniently be grouped under the general heading of intermediary (or intermediate) metabolism. This involves an extensive area of enquiry and for present purposes consideration will be limited to a few selected aspects.

a. Storage Metabolism

The materials, whose breakdown serves for the supply of energy to insects, derive ultimately from ingested food materials which are absorbed from the gut

I.I.P.—2

as soluble molecules. A proportion of these undoubtedly go to satisfy immediate metabolic requirements, but any surplus would be diverted to food depots for storage against a time of need. The main storage depots of insects are the so-called fat bodies, widely distributed in the body and dispersed in cellular sheets between its tissues (see Chapter 4). Since the storage of large quantities of soluble material would raise serious osmotic problems, it is not surprising to find that surplus foodstuffs are stored in cells of the fat body mainly in insoluble form, either as the polysaccharide glycogen or as neutral glyceride or fat.

The formation of glycogen has been shown to involve successive addition of glucosyl units to a preformed glycogen chain through the mediation of a nucleotide complex (see Fig. 1.6 above); and the mobilization of glucose from the polysaccharide involves phosphorolytic cleavage leading to the formation of glucose-1-phosphate which is subsequently converted to trehalose for transport to sites of respiration. It is unfortunate that, despite recent advances in the field of lipid chemistry, there is no comparable information about the synthesis of storage lipids or of the mechanism of their mobilization from storage depots. A certain amount is known of the nature of storage lipids: that they are neutral lipids whose fatty acid constituents comprise both saturated and unsaturated members with chain-lengths predominantly between 14 and 20 carbons. Lipases, capable of hydrolysing the ester linkage between fatty acid and glycerol, have been demonstrated in the fat body of insects, and this reaction presumably constitutes the first step in the mobilization of stored fats. But whether the higher fatty acids enter the haemolymph as such, or whether they are subjected to preliminary degradation first has not yet been determined. In general the biochemistry of lipids in insects stands in urgent need of thorough investigation with modern techniques.

b. Detoxication Metabolism

Carbon dioxide and ammonia have been shown to arise as principal end-products of metabolism (see Fig. 1.1). Both are toxic and require to be removed or detoxicated if malfunction of the metabolic machinery is to be avoided. The elimination of carbon dioxide takes place across the respiratory surface to which it is transported principally as bicarbonate. The initial reaction leading to bicarbonate involves the combination of carbon dioxide with water to form carbonic acid, and in many animals this is catalysed by the enzyme carbonic anhydrase. This enzyme appears to be absent or poorly developed in insects, so that the removal of carbon dioxide does not constitute a metabolic process, and it may best be discussed in connection with the general question of respiratory exchange.

It would appear that the toxicity of ammonia is such as to preclude disposal by simple diffusion in most terrestrial animals. The internal concentration required to ensure a sufficiently high rate of diffusion across relatively impermeable surfaces or along extensive and tortuous diffusion paths of the tracheal system would probably exceed the limits of tolerance, and for this substance it has therefore been necessary to arrange for an alternative method of disposal. This has involved incorporation of the ammonia into a relatively non-toxic molecule, and in insects the principal vehicle for the removal of waste nitrogen is uric acid or substances closely related to it. Like other detoxication products, such as urea, these purines are non-volatile and hence cannot be eliminated by diffusion, and the function of elimination has been taken over by the excretory system (see Chapter 5).

The incorporation of amino acid nitrogen, or of nitrogen deriving from other sources, into the uric acid molecule has been extensively studied in vertebrates, and the synthetic pathway has been fully elucidated. The process is one of considerable complexity, involving a progressive build-up of the purine ring system from molecules or fragments of molecules deriving from quite complex precursors including glutamine, the amide of glutamate, and aspartate. Each step in the synthesis requires the use of ATP as a primer for the reaction, and the expenditure of high energy phosphate may be regarded as the price of detoxication. The derivation of different parts of the uric acid molecule was established by administration of radioactive isotopes of carbon and nitrogen in various forms (e.g. as glycine, formate, bicarbonate etc.) to the synthesizing system, followed by isolation and degradation of the uric acid produced, to establish the position of labelled atoms in the ring system. Figure 1.10 shows the structure of the uric acid molecule and the precursors of its different parts as established in this way.

Comparably exhaustive studies have unfortunately not yet been made of uric acid synthesis in insects. Early work based on the stimulation of uric acid synthesis by administration of possible precursors suggested pathways different from the ones which exist in vertebrates, but the evidence provided by such studies is equivocal. More recent work, based on injection of a radioactive precursor, formate-C^{14}, in the cockroach has shown incorporation of C^{14} in the 2- and in the 8-position, as expected on the basis of the vertebrate pathway. But it is too early to dismiss the possibility that the synthetic system of insects may differ from that of vertebrates in point of detail.

In many insects uric acid is virtually the sole end-product of detoxication (see Chapter 6), but in some there may be substantial degradation of the uric acid molecule by a series of uricolytic enzymes. The first step in this process is the production of allantoin, which involves opening of the imidazole ring under the influence of the enzyme uricase. A second enzyme, allantoinase, may produce cleavage of the pyrimidine ring of allantoin, leading to the formation of allantoic acid. In lower invertebrates degradation may proceed still further, to urea under the influence of allantoicase, and to ammonia under the influence of urease, but there is no evidence that either of these enzymes is active in insects.

It should be mentioned that uric acid may derive from another source than amino nitrogen. Purine nucleotides constitute important degradation products of nucleic acid, and following cleavage of the nucleotide to release the purine base, and deamination of the ring system, the purines can be oxidized to uric acid under the influence of an enzyme, xanthine dehydrogenase. The interrelationship between purine and alpha-amino pathways has been indicated in Fig. 1.10. What proportion of uric acid excreted derives from purine sources will depend to a large extent on the diet of the insect in question. It would be high in insects whose food consists largely of carbohydrates, low in insects whose diets include a lot of protein.

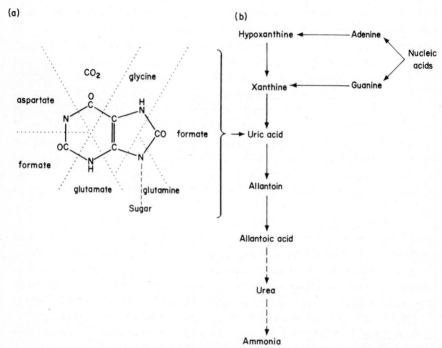

Fig. 1.10. Uric acid and its metabolic relationships. (a) The structure of uric acid in relation to its precursors. (b) Purine degradation.

Urea is a common detoxication product in certain, so-called ureotelic, vertebrates, and this substance can usually be recovered in small quantities from the excreta of insects. It is doubtful, however, whether its occurrence there should be taken as indicating a role in detoxication metabolism. The quantities involved are usually small in relation to the total output of excretory nitrogen, and there is evidence that the ornithine cycle, which is the pathway of its formation in ureotelic animals, is inoperative in the few insects which have been investigated. Further work is necessary to establish the significance of excretory urea and the mode of its formation.

c. *Cuticle Metabolism*

The cuticle of insects functions both as an integument and as an articulated skeleton, and the material of which it is formed needs to embody a corresponding diversity of properties. To serve as a skeleton requires rigidity, yet at points of articulation the cuticle would need to be as flexible as possible. To serve as an integument in a terrestrial animal whose surface area is great in

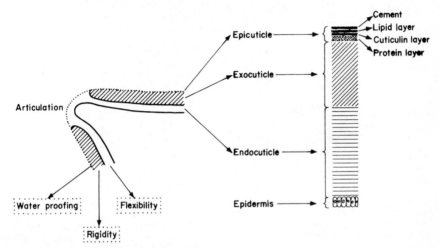

Fig. 1.11. Diagrammatic summary of the structure of the insect cuticle; the scale has been distorted to permit representation of the different layers of the epicuticle.

relation to its volume, water-proofing would be of paramount importance. These different requirements are taken care of by different layers of the composite cuticle, each with its own characteristics (see Fig. 1.11). Flexibility is a property of the relatively thick endocuticle which forms a continuous layer, as does the thin epicuticle that confers the property of impermeability to water; rigidity is a property of the exocuticle, which is discontinuous, and thus provides for points of flexible articulation between rigid regions. The properties of the different layers of the cuticle reflect their chemical composition, and this is summarized below. The way in which the cuticle is formed, through the activity of epidermal cells, will be described in Chapter 13 under the heading of Growth.

(i) The Endocuticle. The endocuticle is a lamellated structure composed principally of microfibrils of protein and of a characteristic polysaccharide, chitin. Recent work by Bouligand (1965) has indicated that the microfibrils are arranged in layered sheets, with the angle of fibril orientation in the tangential

plane changing between successive sheets in such a way as to produce the appearance of parabolically disposed fibres, as typically seen in sections of insect cuticle (see Fig. 1.12). Such an arrangement would allow for distortion in different planes, and thus contribute to the flexibility of the endocuticle. The protein fraction appears to comprise a number of different components, some freely soluble, some linked to the structural framework by relatively weak

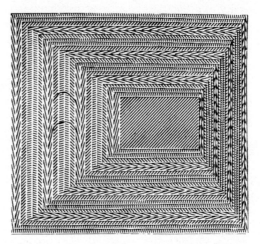

Fig. 1.12. A diagrammatic representation of the derivation of parabolic patterns in cuticle. A truncated pyramid of cuticle is viewed from the top. The chitin-protein microfibrils are arranged parallel to each other forming layers, and the direction of the microfibrils changes from layer to layer. Oblique sections (faces of the pyramid) then produce patterns with apparent parabolic microfibrils, two of which are indicated on the left face (Neville *et al.*, 1969 from Bouligand).

hydrogen bonds and van der Waal's forces, and some combined with chitin forming a glyco-protein complex. Chitin itself is a polysaccharide of high molecular weight, composed of unbranched chains of (1,4)-linked monomers of 2-acetamido-2-deoxy-D-glucose.

$$\text{H}\qquad\text{CH}_3\text{CO.NH}\qquad\text{CH}_2\text{OH}$$

It is likely that a uridine diphosphate complex of N-acetylglucosamine is involved in its synthesis (cf. synthesis of glycogen, p. 20), but little is known of the details of the process in insects.

(ii) The Exocuticle. The exocuticle is a special region of the endocuticle

which has become stabilized and hardened by a process known as sclerotization. The chemistry of this process has been under intensive investigation for the past 20 years, and the general situation is clear although certain details are still imperfectly understood. It is conveniently studied during the formation of the puparium of flies, where extensive sclerotization takes place over a limited period of time, resulting in the transformation of the soft white maggot into the hard and black "pupa". At this time there is a substantial increase in the tyrosine content of the blood, and the activity of the enzyme phenolase, responsible for the oxidation of tyrosine and other phenols to their polyphenol analogues, also shows an increase. The dihydroxyphenylalanine (DOPA) formed by the oxidation of tyrosine (see Fig. 1.13) appears to be deaminated under the influence of enzymes in the blood, in the epidermis or in the cuticle itself, and the polyphenol so formed diffuses to the outer layers of the cuticle where it combines with oxygen in the presence of phenol oxidases, leading to the formation of the corresponding quinones. The quinones now diffuse inwards into the outer layers of the endocuticle, tanning the protein constituents, and converting the flexible chitin/protein complex into a rigid and hard material. The tanning itself appears to involve a reaction of the terminal amino groups of proteins to give N-catechol proteins which, in the presence of excess quinone, are oxidized to the corresponding quinonoid proteins. Further substitution can occur at other points of the benzene ring, and in this way bridges are formed between different protein molecules. These cross-connections are not necessarily confined to the ends of the protein molecules, since it has been shown that the terminal amino groups of lysine residues may become extensively cross-linked, giving stabilization of the whole lamellar complex, and preventing the sorts of deformations which can occur in the untanned cuticle.

Hardening of the cuticle by quinone tanning is usually, though not invariably, associated with darkening. The brown coloration is thought to arise when an excess of quinone is present; under these circumstances the quinones will tend to polymerize to form large pigmented molecules. Such polymers would have a large number of reactive sites, and would be capable of bridging large distances between protein chains.

The main element of controversy in the field of cuticle hardening concerns the precise nature of the tanning agent and the details of its combination with protein residues. A variety of closely related diphenols have been isolated from insect cuticle, but whether some or all of them are indeed involved in the process of tanning has not yet been unequivocally established; it has been said that nowhere in the field of cuticle chemistry has speculation been more popular, and further discussion would clearly fall outside the scope of the present work.

(iii) The Epicuticle. The epicuticle is the outermost layer of the cuticle, which may comprise no more than 5% of the total thickness. What it lacks in dimension, however, the epicuticle makes up in complexity. Four distinct layers

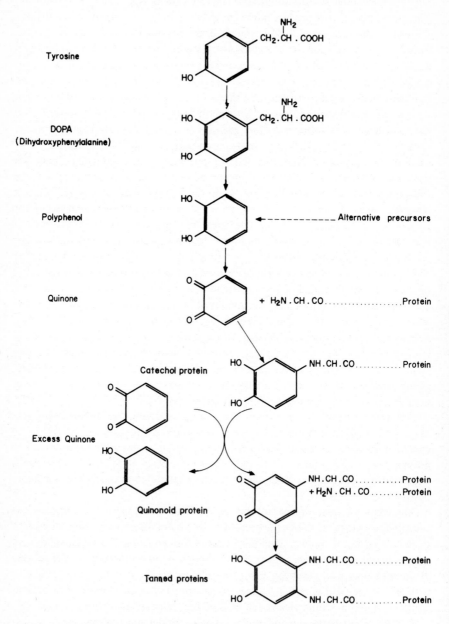

Fig. 1.13. Reactions involved in the hardening of insect cuticle.

can usually be recognized on the basis of suitable staining techniques (see Fig. 1.11):

    (a) an outer layer of resin-like "cement";

    (b) a layer of lipid;

    (c) a cuticulin layer which forms a substrate for the lipid; and

    (d) an inner layer composed apparently of tanned proteins impregnated with lipids, and containing no chitin.

Unfortunately the attenuated nature of the epicuticular layers has precluded detailed investigation of their chemical composition. The only component which has been subjected to rigorous analysis is the lipid fraction. In the insects so far examined the lipids have been found to comprise a complex mixture of different lipid classes, including usually a high proportion of saturated hydrocarbons with chain lengths ranging from C_{12} to C_{30}, and of free fatty acids and alcohols together with their esters. In addition a small proportion of aldehydes and of phospholipids have been recovered. There is good evidence that it is the lipid fraction which confers the property of impermeability to water, but it has not yet been established which of these various components are concerned in water-proofing, or whether different functions may be subserved by different constituents. The relation between cuticular lipids and water permeability will be discussed more fully in Chapter 16.

d. Pigment Metabolism

One of the most striking characteristics of the insects as a group is the diversity of pigmentation shown by some of its members. In many species colour has a physical rather than a chemical basis and depends on the structural conformation of surface elements which may produce interference colours or diffraction effects. But in many the colours result from the deposition of true pigments in the cuticle, or in the underlying hypodermis, and a variety of different kinds of pigment may be involved in the colour pattern of a given species.

(i) Melanins. The brown and black pigments of insects are usually assigned to the general class of "melanins", which as yet permits of no precise chemical characterization. One type of melanin forms when insect haemolymph is exposed to the air. In view of the presence of DOPA in the blood of insects, it seems likely that this black pigment constitutes a polymerized indole-derivative of this substance (see Fig. 1.14). In the cuticle, on the other hand, diphenols are probably deaminated before polymerization occurs, and here a pigment is formed which differs substantially from blood melanin in respect of the absorption spectrum. Finally, it appears that certain derivatives of tryptophan, such as kyneurenin and anthranilic acid, may undergo condensation reactions with quinones, and thus enter into the composition of pigment molecules. Thus a wide range of polymerization products is capable of forming, depending on the

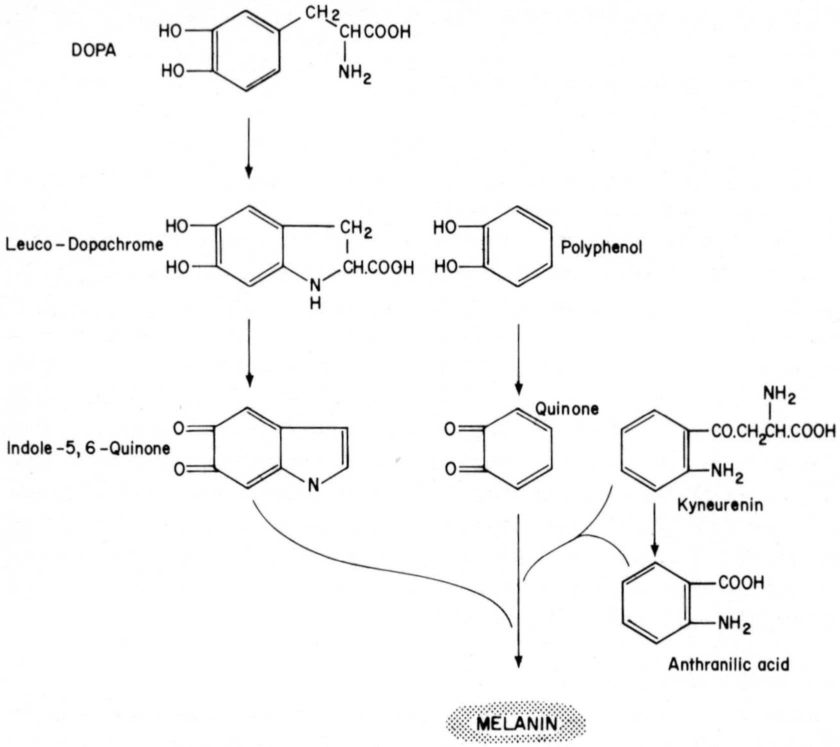

Fig. 1.14. The formation of melanic polymerization products from different precursors.

availability of different types of substrate, and the presence of the appropriate enzymes, the different types having in common their ability to absorb strongly a large range of visible light.

(ii) Pteridines. The pteridines constitute an important group of white, yellow and red pigments in the integument of different species of insect, and most strikingly associated with the epidermal scales of the wings of butterflies and moths. They usually exhibit strong fluorescence, which is a technical convenience, but many of them are photolabile, which has rendered structural investigations a matter of considerable difficulty.

The basic form of pteridine structure is given in Table 1.1, which also shows some of the different pigments produced by substitution at two positions of the ring. Experiments with radioactive tracers have suggested that the pigment may arise from purine precursors, by opening of the imidazole ring and subsequent combination with a dicarbonyl substance like glyoxylic acid.

In addition to their participation in general pigmentation, a number of these pigments have been isolated from the compound eyes of different insects. The

Table 1.1

The structure of pteridines

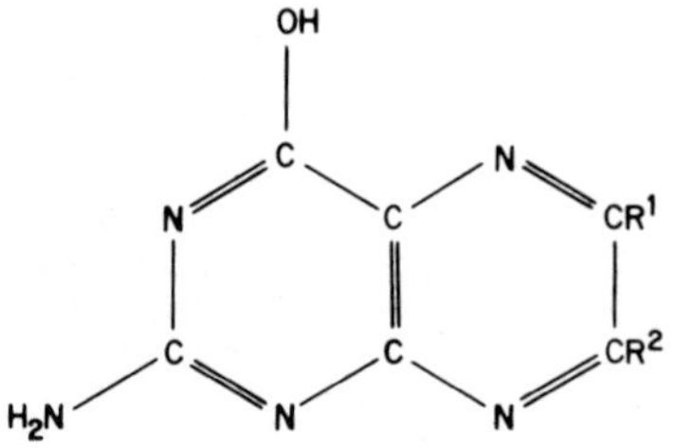

R¹	R²	Pigment	Colour
OH	H	Xanthopterin	Yellow
OH	OH	Leucopterin	White
H	OH	Isoxanthopterin	Colourless
COOH	H	2-amino-hydroxypteridine-6-carboxylic acid	Yellow
OH	CH = C . CH₂ OH $\;\;\;\;$ | $\;\;$ | $\;\;$ OH $\;$ OH	Erythropterin	Red

fact that some of them are extremely labile in the presence of light has raised the possibility that they may play a part in the physiology of vision, but details of such a process have yet to be elucidated.

(iii) Ommochromes. The ommochromes constitute another group of pigments which are involved in general pigmentation as well as being associated with the compound eyes of insects. Their function in vision appears to be confined to processes of adaptation and visual acuity, in so far as they occur in special pigment cells separate from the ommatidia (see Chapter 9). Their migration within these cells in response to changes in light intensity would provide a means of regulating the amount of light which reaches the photoreceptors.

The ommochromes are derivatives of the amino acid tryptophan, an example being the yellow pigment, xanthommatin, which has been isolated from the eyes of blowflies. The compound may be regarded as being a product of an oxidative condensation of two molecules of 3-hydroxykyneurenin, itself a derivative of kyneurenin, which in turn arises by oxidative decarboxylation of tryptophan (see Fig. 1.15).

Other ommochrome pigments, whose precise structure remains uncertain, are also formed from hydroxykyneurenin, as established on the basis of investigations with eye-colour mutants of various insects (see frontispiece). In certain species, for instance, it has been shown that a strain differing from the wild type by a single mutation has reddish-yellow eyes instead of the normal dark

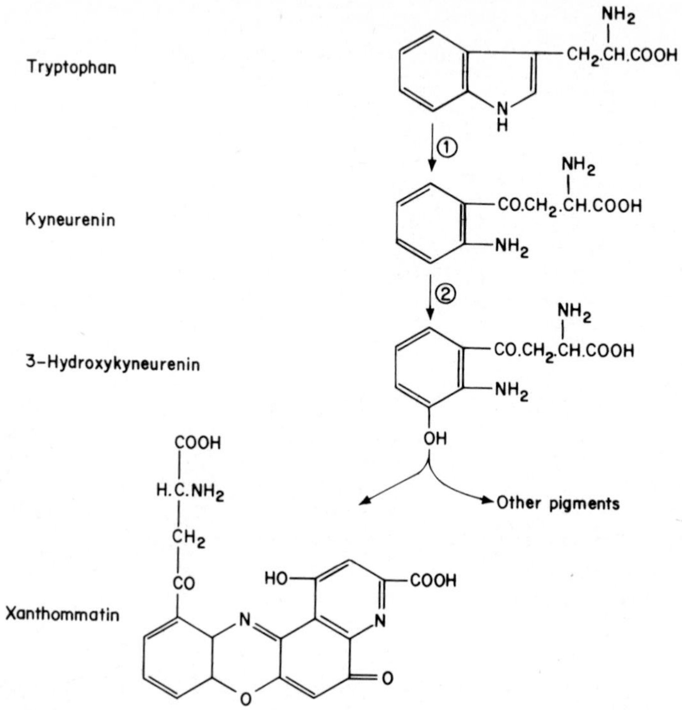

Fig. 1.15. Aspects of ommochrome metabolism. Reaction (1) is controlled by a single gene (v^+) and reaction (2) by another (cn^+).

coloration. Subsequently, it was established that in this mutant the enzyme responsible for the conversion of tryptophan to kyneurenin was inactive; the defect in pigmentation could be overcome by injection of kyneurenin into the developing insect. In another eye-colour mutant, normal pigment synthesis could be restored by the injection of hydroxykyneurenin, but not of kyneurenin, suggesting that here the enzyme responsible for the hydroxylation of kyneurenin was lacking. The indications provided by such studies concerning the pathway of ommochrome metabolism were subsequently verified by injection of radioactive precursors, and the demonstration of radioactivity in the pigments concerned. Thus the pattern of metabolism has been elucidated here by close co-operation between biochemists and geneticists.

CHAPTER 2

THE SUPPLY OF FOOD

A metabolic system like that described in Chapter 1 requires an input of complex organic molecules to serve for the supply of energy, and to provide the raw materials necessary for repair and growth of the system itself; it requires, in other words, an input of food. This food is obtained from the environment, which contains an abundance of organic molecules of many different kinds, and there are few that do not provide sustenance for one or another species of insect. Most of the available food material occurs in the form of organic polymers, large molecules which cannot readily cross the cellular membranes, and are therefore not directly available to the metabolic system. They must be subjected to preliminary manipulation, involving mechanical and enzymatic breakdown, to get them into a form capable of being assimilated into the organism. This is achieved in the process of digestion, which is followed by the absorption of food materials into the body of the animal. But it is not enough that the metabolic system be kept supplied with complex organic molecules, it must be supplied with molecules of the right kind. For purposes of energy release any one of a variety of molecules will serve, but for purposes of maintenance and growth a number of quite specific requirements are involved. For it would seem that in the course of evolution insects, like other heterotrophic organisms which rely for their food on organic molecules elaborated by other organisms, have lost the ability to synthesize a number of essential components of their metabolic systems. In order to maintain their metabolic machinery, and in order to grow and reproduce, they must be supplied with a variety of specific molecules, which they need but have lost the power to make. The supply of food therefore involves a problem of nutrition, of the supply of the right kind of food, and this will be considered in the last section of this chapter.

1. Food Intake

The digestion and absorption of food takes place in the alimentary canal, which in insects is divided into three main parts (see Fig. 2.1(a)): the foregut (or stomodaeum) and the hindgut (or proctodaeum) are derived from ectoderm, and

31

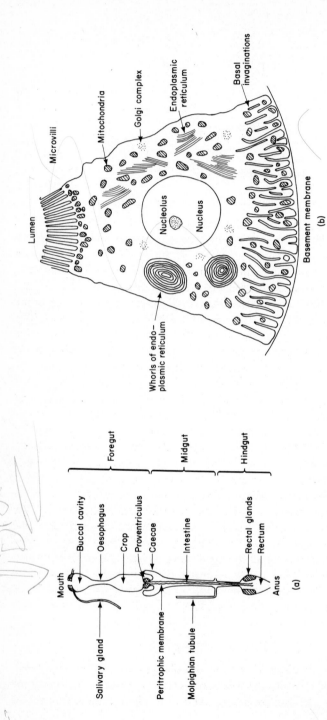

Fig. 2.1. (a) Diagram to illustrate the main features of the alimentary canal in insects. (b) Simplified diagram of a section through a cell of the midgut of the mosquito; on the left of the nucleus the endoplasmic reticulum is represented as it appears in the starved insect, on the right as it appears 24 hr after a blood meal (drawn from electron micrographs of Bertram and Bird, 1961).

are provided with a lining of cuticle; the midgut (or mesenteron) is derived from endoderm. In many insects the midgut is lined by a so-called peritrophic membrane, but this lining is not produced by the cells of the midgut; it represents a secretion from a specialized region called the proventriculus, which marks the junction of foregut and midgut.

The division of the alimentary canal into major regions is usually reflected in a broad differentiation of function. The foregut, with associated structures such as mouthparts and other cuticular elements, is concerned to a large extent with the mechanical breakdown of food material, which may be a necessary preliminary to efficient enzymatic breakdown. It is often possible to distinguish a number of subdivisions of the foregut: a buccal cavity, for instance, which may be furnished with chitinous plates for triturition; a slender oesophagus; and a crop which may serve as a storage organ, and which may be more or less separated from the mainstream of food passage.

The midgut is largely concerned with the secretion of digestive enzymes, the breakdown of polymers under the influence of such enzymes and the absorption of digestive products and of water from the food mass. The ultrastructure of midgut cells is correspondingly complex (see Fig. 2.1(b)); each cell shows the typical folding of basal and peripheral membranes associated with absorptive function; and the endoplasmic reticulum exhibits pronounced cyclical changes, thought to be associated with the synthesis of digestive enzymes and their transport to the lumen. The midgut may carry a number of evaginations (gastric caecae) which serve to increase the surface area available for absorption.

The hindgut receives indigestible remains of the food and stores them pending evacuation. The situation in this region is complicated by the fact that excretory materials from the Malpighian tubules empty into the alimentary canal at the junction of midgut and hindgut, and further processing of the mixture of materials occurs in the rectum as a result of the activity of the rectal glands. Since this would appear to be associated primarily with processes of osmoregulation and excretion, rather than with those of digestion, it will be dealt with in another chapter (see Chapter 5).

The functional distinction which may be drawn between different regions of the alimentary canal is by no means absolute. Salivary glands, which may produce digestive enzymes, often open into the buccal cavity, so that a certain amount of digestion takes place in the foregut of many species in addition to the mechanical breakdown. Indeed in some insects saliva may be extruded on to the food mass before it enters the alimentary canal. On the other hand, a substantial amount of mechanical breakdown undoubtedly occurs as a result of the peristaltic contractions of the midgut, although the primary function of these contractions may be the onward propulsion of the food mass. Finally, it has been found that absorption of digestive products may occur in the hindgut as well as in the midgut of certain species of insect. Even so, the general organization of the alimentary canal shows a surprising degree of uniformity in different species of insect. Variations on the common theme usually involve such things as the occurrence of a crop as a separate blind diverticulum of the foregut, and the presence or absence, and degree of development, of midgut caecae. It is only in the specialization of mouthparts that the sort of diversity normally associated with the insects as a whole is encountered. Arrangements range from the relatively simple grinding or shearing mouthparts of insects like the cockroach and the locust to the highly specialized adaptations shown by the blood-sucking mosquitoes or the nectar-feeding butterflies. A discussion of the details of such special adaptations would, however, lie outside the scope of the present work.

2. Digestive Enzymes

The bulk of food material available to insects is constituted by the three main classes of foodstuff—carbohydrate, protein and fat, in proportions depending on the diet of the different species. A corresponding diversity of digestive enzymes, capable of hydrolysing these different polymers to their constituent units, will therefore be required by most insects. Some of these enzymes are elaborated in special salivary glands and discharged into the buccal cavity, but most are produced by the cells of the midgut.

a. Carbohydrases

A great deal of work has been done to demonstrate the presence of different kinds of enzyme in the alimentary canal of insects, but investigations have seldom proceeded to the point of enzyme purification, nor have the conditions necessary for optimal activity been carefully established. Normally, crude extracts of digestive tract are tested for ability to hydrolyse specific substrates, maltose or trehalose for example, and if activity is detected, the presence of the corresponding enzyme, maltase or trehalase, is inferred. But what may be present is an enzyme capable of hydrolysing the α-glucosidic linkage generally, rather than maltose and trehalose specifically. Such results cannot, therefore, be accepted as evidence for the existence of two different enzymes; this could be established only by further work involving, perhaps, electrophoretic separation of protein components in the extract, and demonstration that hydrolysis of the two substrates is associated with different regions of the electrophoretic plate. In the absence of such data, and in the interests of simplicity, it will, for present purposes, be convenient to distinguish only a few general classes of carbohydrases.

(i) Glucosidases. Enzymes capable of hydrolysing the α-glucosidic linkage (as in maltose and trehalose) and the β-glucosidic linkage (as in cellobiose and gentiobiose) appear to be distributed widely among insects. α-Glucosidases are particularly active in herbivorous species, in accord with the fact that most of the naturally occurring α-glucosides are of plant origin. The digestive enzyme which is active towards trehalose is probably different from the completely specific trehalase which has been discussed in the chapter on metabolism.

(ii) Galactosidases. Enzymes capable of hydrolysing the α-galactosidic linkage (as in melibiose and raffinose) have also been demonstrated in a range of insect species. β-Galactosidases, on the other hand, which would be required for the hydrolysis of carbohydrates like lactose, have been demonstrated in only a few species, of which the mealworm beetle is one.

(iii) Fructosidases. Sucrose is a β-fructoside, and enzymes capable of hydrolysing it have been isolated from the alimentary canal of a number of species of fly, but they do not appear to be widely distributed among insects generally.

(iv) Amylases. These enzymes split the storage polysaccharides of animals and plants (glycogen and starch respectively) to the disaccharide maltose, which is further degraded under the influence of α-glucosidases. Active preparations, differing considerably in respect of pH optimum which ranges from 5.5 to 9.5, have been isolated from the alimentary canal of a wide variety of insects.

(v) Cellulases. Cellulase activity has been demonstrated in a few species of wood-boring beetle, and in certain silverfish. Cellulose forms a substantial part of the diet in many insects, but this particular polysaccharide of glucose appears to be extremely resistant to hydrolysis, and most insects have to rely for its digestion on the presence of symbiotic microorganisms in the alimentary canal. Two other intractable polysaccharides, chitin and lichenose, are also capable of being digested by certain insects.

b. Proteinases

The digestion of protein in insects appears to follow the same general pattern as in other animals, with a preliminary breakdown of the large protein molecules, followed by an attack on the smaller peptones and polypeptides so formed, which are ultimately hydrolysed to their constituent amino acids; and in insects, as in other animals, a distinction can be made between different types of peptidase, some attacking from the carboxyl end, some from the amino end of the peptide, and some responsible for the hydrolysis of the dipeptide end-products of the previous attacks. Insect proteinases have been found generally to be more active in neutral or alkaline media than in acid, and such work as has been done on isolation and purification suggests that a number of different enzymes may be involved in the first stages of protein breakdown.

Of particular interest is the occurrence in certain insects of enzymes capable of attacking relatively resistant materials such as collagen and keratin. In keratin, the protein molecules are cross-linked by disulphide bridges associated with the cystine residues, and the ability to break the interconnections appears to be limited to a very few species of insect, belonging to quite different orders—the clothes moth, a dermestid beetle and certain biting lice of birds. The reduction of the disulphide bonds effected by these insects appears to destroy the resistance of keratin to enzymatic attack and enables breakdown of the protein molecule by normal proteolytic enzymes.

c. Lipases

Enzymes capable of hydrolysing neutral fat to glycerol and free fatty acids have been demonstrated in a wide variety of insects, though not all the species examined have yielded active extracts. Most of the lipases concerned show maximum activity at alkaline pH values, and under these conditions the fatty acids released would form soaps, which would help to emulsify undigested fat, and thus facilitate further action of the hydrolytic enzymes.

The diet of the wax moth, *Galleria mellonella*, consists predominantly of beeswax, and the mechanism of digestion of this intractable material has been the subject of a number of investigations, but the situation has not yet been satisfactorily elucidated. It is thought that symbiotic microorganisms may co-operate in the process, which appears to be associated with an active phosphorous metabolism, but the significance of this association remains obscure.

This brief review of the digestive enzymes of insects has served to indicate that members of the class have the enzyme complement to deal with all of the three main classes of foodstuff, and that a number of species have developed special adaptations to special feeding habits which enable them to digest animal products, such as chitin, keratin, collagen and beeswax, and plant products such as cellulose and lichenose, which would normally be considered highly resistant to digestive attack. Over and above these special adaptations, it is often possible to see some correlation between general enzyme endowment and feeding habit. For instance, in carnivorous insects, such as certain species of blowfly, proteinases and lipases are strongly developed, while carbohydrase activity is feeble. Conversely, in plant-feeding beetles like *Melolontha* carbohydrases predominate over proteinases. But the diet of such insects, however narrow their food preference, is not completely restricted to a single class of food material, and the distinctions which can be drawn are only quantitative. Even in obligatory blood-suckers, like the tsetse fly, whose food contains negligible proportions of polysaccharide, digestive juices show weak amylase activity.

3. The Control of Digestion

Studies of digestion have produced evidence that the production of digestive enzymes may be under control of the neuroendocrine system in many insects, particularly in those whose feeding is markedly discontinuous. The release of hormones which stimulate the production of enzymes may be to some extent associated with the normal process of development, as in *Tenebrio*, the meal

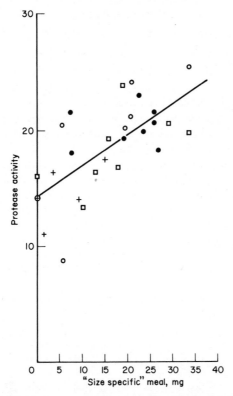

Fig. 2.2. The relationship between midgut protease activity and the size of the blood meal in the tsetse fly. ●, whole blood; □, 50% blood; ○, 25% blood; +, 10% blood (Langley, 1966).

worm beetle, where proteinase production is triggered by neuroendocrine release at emergence from the pupa. But in addition to such general control, there may be a process of regulation in relation to the act of feeding and to the amount of food ingested at feeding. In blood-sucking insects like black-flies and tsetse flies, feeding stimulates synthesis of proteolytic enzymes by cells of the midgut. In the tsetse fly the amount of enzyme produced, and hence the rate at which digestion takes place, is closely related to the size of the blood meal (see Fig. 2.2). The effect can be demonstrated whether the fly is fed on whole blood or on various dilutions of whole blood, suggesting that it is the volume of the meal, rather than the amount of protein it contains which is the important factor. It appears that the correlation between meal size and proteinase activity is mediated by stretch receptors situated in the wall of the crop and perhaps elsewhere, which monitor the degree of distension of parts of the alimentary canal.

4. Absorption

The absorption of digestive products from the lumen of the midgut has not been extensively studied in insects; discussion of the problem will therefore be based largely on a detailed series of investigations of the process as it occurs in the locust and the cockroach. In these, as in several other insects, the rate of absorption of food materials is controlled by the rate at which food is released from the crop, which in turn depends on the concentration of food in the contents of the crop, the volume of food material leaving it decreasing with increasing concentration. In order to study the process of absorption, it was necessary to eliminate this element of control, and this was done by injecting the food material into the alimentary canal by way of the hindgut. Changes in the concentration of material caused by absorption of water were established by incorporation in the food solution of the dye amaranth, a substance which is not absorbed by the insects under investigation. The use of radioactively labelled carbohydrate, amino acid and fat enabled ready determination of their concentration in the gut contents and in the haemolymph.

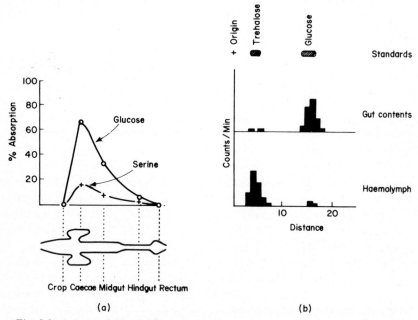

Fig. 2.3. Aspects of digestion in insects. (a) The percentage of material absorbed from different parts of the alimentary canal of the locust 15 min after the injection of glucose (20 mM) and of serine (35 mM). (b) Radioactivity on paper chromatograms of samples of the caeca contents and of haemolymph 15 min after injection of C^{14}-labelled glucose (20 mM) into the alimentary canal. The position of trehalose and of glucose standards is shown at the top of the figure. Data from Treherne (1958) and (1959).

Selected results of these investigations are illustrated in Fig. 2.3(a), which shows that the rate of absorption, both of glucose and of the amino acid serine, is greatest in the midgut caecae, appreciable in the midgut itself, and that little absorption takes place in the foregut or in the hindgut.

The process of absorption was found to be associated with an increase in the concentration of food materials in the alimentary canal, brought about by a withdrawal of water from the injected solution. Radioactive material in the gut lumen exchanged freely with unlabelled material in the haemolymph, but a net transfer of food from the gut to haemolymph was imposed by the concentration gradient associated with the higher gut concentration.

The rate of transfer of glucose was very much higher than the rate of transfer of amino acids. This difference was found to be correlated with the transformation of glucose to trehalose during passage across the gut wall, as illustrated in Fig. 2.3(b). The effect of this coupling of two molecules to form trehalose on the haemolymph side of the gut wall would be to maintain a low concentration of glucose in the haemolymph, and hence to ensure that a steep concentration gradient is maintained between gut and haemolymph, favourable for the rapid absorption of glucose. The larger size of the trehalose molecule, and its relatively low molar concentration, would to some extent reduce the back-diffusion of trehalose into the gut, so that the over-all effect would be a "facilitated" uptake of glucose.

The haemolymph concentration of amino acids tends to be high, hence the concentration gradient available for the transfer of material from gut to haemolymph remains relatively shallow. A diagrammatic summary of the situation is shown in Fig. 2.4.

It is possible that the occurrence of trehalose as the principal blood sugar in insects should be seen in part as a reflection of this mechanism of facilitated diffusion. Glucose constitutes a major dietary constituent for the majority of insects, and its uptake from the gut would be greatly hindered by the occurrence of high concentrations of glucose in the haemolymph. On the other hand, a high concentration of blood sugar is a primary requisite for the rapid transfer of substrate to actively respiring tissues generally, and to flight muscle in particular (see Chapter 1). These conflicting requirements are resolved by the interpolation of a secondary blood sugar, which enables a favourable concentration gradient to be maintained both for the transfer of glucose from gut to haemolymph, and for the transfer of substrate from haemolymph to the site of respiration. It may be that an analogous situation exists in the tsetse fly and perhaps in other insects that subsist on high protein diets. Here the main haemolymph amino acid, which may serve as a respiratory substrate, is proline; this substance is poorly represented in the food intake compared with substances like glutamic acid, aspartic acid, alanine and glycine, all of which stand in even closer relationship to energy-yielding pathways. The occurrence of high concentrations of proline in

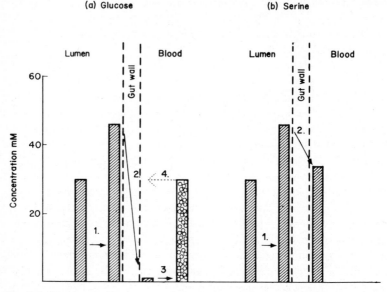

Fig. 2.4. Diagram to illustrate the factors involved in the absorption of food materials from the midgut of the locust. (a) Glucose injected at a concentration of 30 mM; (1), increase in concentration due to absorption of water giving (2) a steep diffusion gradient across the gut wall; absorbed glucose is converted to trehalose (3), and back-diffusion of trehalose (4) restricted by large molecular size and relatively low molarity. (b) Serine injected at a concentration of 30 mM: (1) increase in concentration due to absorption of water, but the diffusion gradient is shallow (2) because of the high haemolymph concentration, and serine is therefore absorbed at a much slower rate than glucose. Data from Treherne (1958) and (1959).

the haemolymph would therefore present no obstacle to the rapid uptake of the bulk of digestive products from the gut.

The absorption of fat has not been investigated so extensively, but it has been shown that here again the midgut caecae are responsible for the bulk of absorption. In the cockroach the rate of uptake is greater with partially hydrolysed material, but total degradation does not appear to be essential for absorption.

5. Nutrition

An intake of carbohydrate, or of protein or of fat, will furnish the metabolic system with its requirement in relation to the provision of energy. Any one of these materials can be oxidized through the Krebs cycle, and a part of their energy trapped in high energy phosphate linkage for subsequent use by the system. But more than energy is required for the maintenance of the metabolic system; enzymes need to be synthesized, co-factors of various kinds must be made available, and a variety of highly complex molecules must be elaborated to serve as components of intracellular and extracellular architecture. Many of the compounds needed for these purposes can be furnished from simple starting materials through normal pathways of intermediary metabolism, but there is a large number which insects, in common with most other animals, are incapable of synthesizing, and these must be provided ready made as elements of diet.

The need for specific raw materials of this kind is associated with the general turnover of tissue components, which has been shown to characterize living organisms in general, and it is therefore ever present. It becomes acute, however, during periods of growth and during periods of reproduction, particularly in the female who must make the more substantial contribution to the resources of developmental stages. Nutritional requirements are therefore usually assessed on the basis of the capacity of specific diets to maintain growth and reproduction. One of the main difficulties in making accurate assessments on this basis arises from the presence of different sorts of microorganism in the alimentary canal of most insects. As has been mentioned above, these may play a specific part in the digestion of particular types of food, but apart from this they will serve as a source of raw materials of the kind under discussion, by virtue of their ability to synthesize a much greater variety of organic molecules than their hosts. These molecules will become available to the host in the course of the normal turnover of populations of microorganisms, being released by the death and disintegration of organisms within the alimentary canal. It may therefore be possible for an insect to sustain normal growth and reproduction on a diet which in reality is deficient in certain respects, thanks to the supplement derived from its intestinal fauna. To obtain unequivocal evidence of the importance of a given constituent of diet it would be necessary to work with insects which have been deprived artificially of their intestinal microorganisms and to achieve this is usually a matter of considerable technical difficulty. For this reason the number of insects for which nutritional requirements have been firmly established remains small.

Nutritional requirements of three different kinds are usefully distinguished:

(a) the requirement for specific amino acids, which would be needed for the elaboration of enzymes and other proteins. Since proteins are a major constituent of the living organism, the need for raw material in this category is substantial, usually of the order of milligrams per gram dry weight of diet:

(b) the requirement for certain lipids, including unsaturated fatty acids and steroids, which constitute important constituents of cellular membranes. The quantities required are usually much less than a milligram per gram dry weight of diet; and

(c) the requirement for vitamins. Many of these have been identified as co-factors and enzyme components, and are needed, therefore, only in catalytic amounts. Adequate provision for growth is made at a level of micrograms per gram dry weight of diet.

a. Amino Acids

Amino acids are distinguished as essential or non-essential, depending on whether they need to be provided in the diet, or whether they can be synthesized from non-nitrogenous sources. The essential amino acids in insects are arginine, histidine, isoleucine, leucine, lysine, methionine, phenylalanine, threonine, tryptophan and valine, the same 10 as in the rat and many other mammals.

Non-essential amino acids include alanine, aspartic acid, glutamic acid and glycine, not surprisingly in view of the fact that the keto-analogues of these amino acids are members of the pathway of carbohydrate degradation (see Chapter 1), so that they can be synthesized from carbohydrate precursors by the simple process of transamination. Proline and hydroxyproline, which can be derived from glutamic acid, also belong to this category, as does serine, which appears to be capable of derivation from glycine. Tyrosine, and the sulphur-containing amino acid cysteine, are not essential provided their essential precursors, phenylalanine and methionine, are present in excess of minimal requirements.

b. Lipids

Insects appear to be unable to synthesize steroids, and they therefore show a specific requirement for cholesterol, or for some suitable precursor in the form of a short-chain derivative like ergosterol, which can substitute in most species.

Cholesterol

Other derivatives, like stigmasterol or sistosterol, which are characteristic of plant tissues, may satisfy the requirements of phytophagous species, but cannot usually do so in carnivores.

Many insects appear to synthesize all fatty acid requirements from non-lipid sources, and in fact the inclusion of fatty acids in the diet often appears to be detrimental. In a few species, however, a requirement for certain unsaturated fatty acids has been demonstrated (e.g. linoleic acid for Lepidoptera).

c. Vitamins

The study of vitamin requirements is complicated by the fact that such very small quantities are involved; sufficient amounts may therefore be held in reserve to tide the insect over periods of temporary deprivation, and in some species deficiency symptoms may not appear for several generations. Under such circumstances, the use of antivitamins, substances which inhibit the activities of specific vitamins, may provide more reliable evidence of requirements.

The following vitamins have been found to be essential dietary constituents for all species which it has been possible to study uncontaminated by microorganisms.

(i) Thiamine. A constituent of enzymes involved in carboxylation and decarboxylation.

(ii) Riboflavin. A constituent of the flavoproteins, which play a part in the transport of hydrogen.

(iii) Nicotinic Acid. A constituent of the coenzymes of dehydrogenation.

(iv) Pantothenic Acid. A constituent of coenzyme *A*.

(v) Biotin. The precise nature of its action is obscure, but this compound is thought to be involved in reactions of deamination and decarboxylation.

(vi) Choline. This serves as a donor of methyl groups and is a constituent of acetyl choline and of certain phospholipids.

The following have been found to constitute dietary requirements in most insects, but certain species appear to be capable of growth and reproduction without them.

(i) Pyridoxine. A coenzyme of transamination.

(ii) Folic Acid. A substituted pteridine which serves as a coenzyme in group transfer reactions involving formate.

The following vitamins are apparently synthesized by most species of insect, and are not required as dietary constituents.

(i) Carnitine. A substance involved as a transport factor in the oxidation of fatty acids (certain tenebrionid beetles are exceptional in showing a requirement for this substance).

(ii) Fat-soluble Vitamins. These include vitamins A, D, E and K.

6. Conclusion

The supply of food to the metabolic machinery of insects appears to show few features which could be regarded as typical of the class; within the group there have been many adaptations to different feeding habits, but the general way in which the food materials are manipulated once they have been ingested shows no striking peculiarities, and the requirement for specific items of diet is very similar to that of other animals. The only feature which appears to be specifically related to the insectan organization is the "facilitated" absorption of glucose.

CHAPTER 3

THE CIRCULATORY SYSTEM

When food materials have been absorbed through the walls of the alimentary canal they are transported by the circulatory system to sites of tissue respiration and synthesis, or to sites of food storage. It will therefore be appropriate at this stage to consider the nature of the circulating medium and the mechanism of its circulation. First it should be mentioned that the extracellular fluid under consideration serves not only as a circulating medium, analogous to the blood of higher forms, but also as a bathing medium for the cells of many tissues, analogous to the lymph of vertebrates. By reason of this dual function it has been given the name of haemolymph.

1. The Composition of Haemolymph

a. Soluble Components

The concentration of soluble components in insect haemolymph is extraordinarily variable. Freezing point depressions, which provide a convenient measure of total osmotic concentration, range from $0.4°$ to $1.1°$ in different insects, corresponding to a range of between 0.7% and 1.9% sodium chloride equivalents. Differences are substantial not only between species, but between different stages in the life history of a single species, and even within a single stage, depending on variations in physiological state.

Even more striking than the variation in total osmotic concentration is the variation in proportionate composition, examples of which are shown in Fig. 3.1. In many species, particularly among the primitive classes (Apterygota and Exopterygota), inorganic ions make up the bulk of osmotically active material (Fig. 3.1(a)), with sodium as the predominant cation and chloride as the predominant anion. Members of the Phasmidae are peculiar in this respect, with very high concentrations of magnesium and phosphate (Fig. 3.1(b)). In the more advanced classes, among the Endopterygota, inorganic ions constitute a much smaller proportion of the total osmolar concentration (Fig. 3.1(c)), and they make up only about 25% in many butterflies and moths (Fig. 3.1(d)) and even less in certain beetles (Fig. 3.1(e)). In such insects the bulk of osmotic activity is

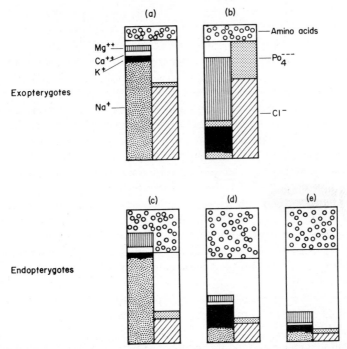

Fig. 3.1. Proportionate ionic composition of the haemolymph of different insects. The blocks represent total osmolar concentration, with relative cation contributions shown on the left and relative anion contributions on the right. The large blank areas represent the proportion accounted for by components of the haemolymph other than those listed (redrawn from Sutcliffe, 1963). (a) Various exopterygote orders (e.g. Odonata, Plecoptera, Heteroptera). (b) Phasmidae. (c) Various endopterygote orders (e.g. Neuroptera, Trichoptera, Diptera). (d) Lepidoptera and Hymenoptera. (e) Certain Coleoptera.

accounted for by amino acids and other organic materials. Here members of the Lepidoptera and Hymenoptera are of interest in having relatively high concentrations of potassium, which often substantially exceed the sodium concentrations (Fig. 3.1(d)). It was thought that this characteristic might be associated with the plant feeding habit which characterizes these classes, since plant material tends to be relatively rich in potassium. But, in fact, many phytophagus species have been shown to have high proportions of sodium, and other factors clearly come into play to determine the sodium/potassium ratio. Whatever their cause, these variations in the proportion of inorganic ions are of considerable interest in relation to the functioning of excitable tissues, and they will be discussed further in that context (see Chapter 8).

In most insects a substantial fraction of the osmotic activity is made up by free amino acids, which occur in concentrations ranging from 200-2000 mg/100 ml. A great deal of work has been done to elucidate quantitative details of the pattern in different species and these studies have revealed a situation of bewildering complexity. Most amino acids have been shown to be present in all the insects studied, but their relative importance is subject to spectacular variation. Virtually any amino acid may constitute a major element in the haemolymph of one species at some stage in its life history, and be virtually lacking in another species. Part of the recorded variation is probably associated with short-term fluctuations in the concentration of particular amino acids associated with metabolic activities of different kinds. In the silkworm, for instance, the concentration of methionine, glutamate and aspartate fluctuates greatly in the course of larval and pupal development, correlated with changes in the secretory activity of the silk glands. Similarly, in the tsetse fly, the concentration of proline may be extremely high in the resting insect but it decreases greatly during flight, while alanine shows the converse relationship. Instances of this sort suggest that a satisfactory interpretation of the amino acid pattern in insects must await elucidation of the part which different amino acids play in the metabolism of the different species. In the absence of such information, only the broadest and most tentative generalizations can be hazarded, namely:

(a) that certain amino acids including aspartic acid (and its amide asparagine), leucine and isoleucine tend to be poorly represented in the haemolymph of insects; and

(b) that glutamic acid (and its amide, glutamine) and proline are usually well represented in the haemolymph of insects, and often attain levels of concentration equal to or exceeding those of trehalose. It is possible that this should be seen as a reflection of the part played by both of these substances as substrates for flight metabolism as discussed in Chapter 1.

Another major organic constituent of insect haemolymph is trehalose, whose function has already been discussed. In the resting insect concentrations normally range from 500-5000 mg/100 ml depending on the species, thus constituting a substantial fraction of the osmolar concentration. Glucose and other carbohydrates are generally present in very much lower concentration, an exception being the honey bee, where both glucose and fructose concentrations may be high. Another carbohydrate, glycerol, has been demonstrated in extremely high concentration in the haemolymph of a number of species; its occurrence in quantity is usually associated with the development of cold-hardiness, and the subject will be discussed further in Chapter 15.

Other substances which may contribute substantially to osmotic activity in the haemolymph are a variety of organic acids of the Krebs cycle, some of which may attain to concentrations of 200 mg/100 ml, and play a major part in ionic balance, accounting for as much as 40% of cation binding. Organic phosphates, such as α-glyceraldehyde phosphate and glucose-6-phosphate may also be present in quite high concentrations, of the order of 50 mg/100 ml; and nitrogenous waste products, such as uric acid and allantoin, may contribute to osmotic activity at concentrations of 2-20 mg/100 ml.

In addition to these small organic molecules, insect haemolymph contains considerable amounts of soluble protein, generally in the region of 1000-5000 mg/100 ml. With the development of microtechniques for starch gel electrophoresis and similar fractionation procedures, the haemolymph proteins of a number of species have come under investigation. A large number of components can usually be identified, and spectacular changes have been found to occur at different stages of the life history, with particular protein bands appearing at particular stages of development. The functional significance of these changes have not yet been elucidated, except in so far as some of the bands have been shown to be associated with enzymatic activity. Indeed, haemolymph enzymes appear to constitute a surprisingly high proportion of total protein, and many different kinds of catalytic activity have been shown to be involved, including hydrolysis, dephosphorylation, dehydrogenation, oxidation and transamination.

TABLE 3.1

A comparison between the content of soluble material in the haemolymph of insects and in the blood of vertebrates

Class of substance	mg/100 ml	
	Insect haemolymph	Vertebrate blood
Inorganic	750	850
Organic		
proteins	3000	20,000
carbohydrates	2000	90
amino acids	1200	60
organic acids	100	30
sugar phosphates	50	30
nitrogenous waste	10	30

Data for vertebrate blood from "Biochemists Handbook" (C. Long, ed.). E. and F. N. Spon Ltd., 1961.

Bearing in mind the tremendous variation in composition which characterizes the haemolymph of insects, a rough attempt has been made in Table 3.1 to compare the distribution of soluble constituents in the haemolymph of an endopterygote type of insect with that of vertebrate blood. Apart from the very high protein content of vertebrate blood, associated with the development of a respiratory pigment, the most striking difference concerns the concentrations of carbohydrates and free amino acids, which are about 20 times as great in insect haemolymph as in vertebrate blood. The possible significance of this difference will be discussed in greater detail below.

b. Haemocytes

The haemolymph of insects usually contains a variety of cellular elements, collectively known as haemocytes. They do not make up a constant feature of the circulating medium, since at times most of them may form transient aggregations on organ surfaces, leaving few free in the circulation. In some species the haemocytes may even form stable cell masses in specific regions of the body, apparently playing a part in haemotopoiesis, the formation of blood cells. At other times, particularly after injury or during moulting, the major proportion of haemocytes may circulate freely in the haemolymph, the number of cells reaching levels greater than 50,000 per cubic millimetre.

A number of different types of haemocyte, some of which are illustrated in Fig. 3.2 have been described. They may be distinguished on the basis of their shape, the type of nucleus and the presence or absence of cytoplasmic inclusions. The relationship between the various types of cells has not yet been established unequivocally. They appear to perform a variety of functions including:

(i) *Phagocytosis.* Certain types of haemocyte are particularly active in the ingestion of particulate material, whether exogenous (insoluble dyes or dead bacilli injected into the haemolymph) or endogenous (fragments of tissue resulting from histolytic activity). In some insects these phagocytic haemocytes may form loose aggregations which serve as haemolymph filters.

(ii) *Encapsulation.* Haemocytes have been shown to play a part in the encystment of various foreign objects, including metazoan parasites. They form a capsule round the invading organism, and may cause its death by cutting off supplies of food or of oxygen. The encapsulating haemocytes often transform to connective tissue membranes, and a process of melanization sometimes is involved.

(iii) *Wound Healing.* Haemocytes often tend to aggregate at the site of injury, where they may proliferate to form a plug which helps to seal the wound.

(iv) *Coagulation.* The haemolymph of different insects appears to differ considerably in its capacity to form a clot. In some species coagulation seems not to occur at all, in others an apparent coagulation may result from an aggregation of haemocytes, while in many a true coagulation of the plasma occurs, involving the precipitation of haemolymph proteins. A special type of "hyaline" haemocyte appears to play a role in plasma precipitation; disruption of these cells in contact with a foreign surface is apparently associated with the release of a factor which promotes coagulation.

(v) *Metabolism.* Haemocytes appear to play an active part in the intermediary metabolism of insects, as evidenced by the occurrence within them of a variety of stored materials, including glycogen, neutral mucopolysaccharides, phospholipids and proteins. They have been implicated in the formation of basement membranes, of particular hormones, and of melanin.

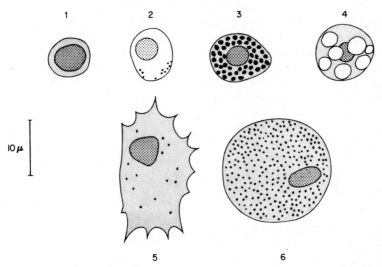

Fig. 3.2. Diagrams of the major types of haemocyte. 1, prohaemocyte, considered to represent the stem cell for other types; 2, coagulocyte, whose disintegration is associated with plasmacoagulation; 3, spherule cell, with large granular inclusions which may contain tyrosinase; 4, adipohaemocyte, with lipid droplets in the cytoplasm; 5, plasmatocyte, a phagocytic haemocyte; 6, oenocytoids, large cells often of irregular shape with granular or crystalloid inclusions (drawn from phase contrast photographs of Jones, 1964).

2. The Pericardial Cells

Before proceeding to a consideration of haemolymph circulation it will be convenient to deal briefly with a type of cell, the so-called nephrocyte, whose functions appear to be essentially similar to those of certain haemocytes. These cells are often closely associated with the dorsal vessel of the circulatory system and under these circumstances they are usually referred to as pericardial cells. They occupy fairly constant stations, as lobes or clusters of cells, on the outer wall of the dorsal vessel or within its lumen. The cells are often filled with red, yellow, green or brown pigments, and their most striking attribute is their ability to take up and store a variety of injected dye materials. It is likely, however that their functions involve more than just a simple segregation of haemolymph contaminants and waste products, especially since the content of such materials shows no progressive increase with age. Like some of the haemocytes, these cells may be active in intermediary metabolism, and they seem also to represent a site of hormone synthesis, as will be shown below.

3. The Circulation of Haemolymph

The haemolymph of insects circulates through a system of sinuses, collectively referred to as the haemocoele, interposed between the various tissues of the body; these haemolymph channels lack a true endothelial lining, and are often rather poorly defined. The main propulsive force is provided by peristaltic contractions of the dorsal longitudinal vessel; in many species its action is reinforced by the activity of accessory pulsatile organs, and by contractions of dorsal and ventral fibromuscular diaphragms (see Fig. 3.3).

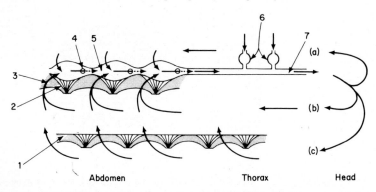

Fig. 3.3. Lateral diagrammatic view of the haemolymph circulation in an insect. 1, ventral diaphragm; 2, alary muscles; 3, dorsal diaphragm; 4, ostium; 5, heart; 6, accessory pulsatile organs facilitating the return of haemolymph from the wings; 7, aorta. (a) pericardial system; (b) perivisceral system; (c) perineural system of haemolymph channels. The heavy arrows indicate the normal course of haemolymph flow (based on Wigglesworth, 1965).

The dorsal vessel is usually a straight tube extending the length of the body, and comprising two main regions:

(a) a posterior "heart" with paired ostia opening into it in each body segment; the ostia are guarded by valves, and the inward projection of these valves may serve to sub-divide the heart into a series of chambers; and

(b) an anterior "aorta" which lacks lateral openings.

The dorsal vessel is suspended from the body wall by elastic filaments, and these serve also to attach it to the dorsal diaphragm when present. This septum underlies the heart, and is usually associated with a series of fan-shaped alary muscles, each converging to an insertion on the lateral body wall. Another diaphragm may be present immediately above the ventral nerve cord. Such transverse membranes tend to divide the system of haemolymph channels into three main regions—a pericardial region surrounding the heart, a perineural region surrounding the ventral nerve cord, and a perivisceral region surrounding the gut (see Fig. 3.3 (a), (b) and (c)). In many species, however, the diaphragm may be extensively fenestrated and would then provide only a partial barrier to the exchange of haemolymph between regions.

A number of accessory pulsatile organs have been described in different insects. They are situated usually at the base of antennae, wings (see Fig. 3.3) or legs, and serve to promote the movement of haemolymph in the corresponding appendages remote from the main pathways of circulation.

Peristaltic contractions of muscles in the wall of the longitudinal vessel, usually initiated at the posterior end and progressing towards the head, serve to force blood into the aorta, from which it is discharged into the anterior sinus systems. From here the blood percolates backwards between the tissues of the body, and is ultimately drawn back into the posterior parts of the dorsal vessel by the fall in pressure associated with its relaxation. The sequence of mechanical events during a cycle of contraction is illustrated in Fig. 3.4(a), which shows a

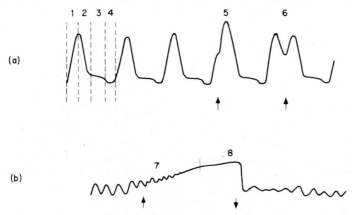

Fig. 3.4. Records of the mechanical activity of the cockroach heart. (a) Normal heart beat, showing systole (1), diastole (2), diastasis (3) and "pre-systolic notch" (4); at (5) the heart was stimulated with a single shock during systole, as marked by the arrow, and at (6) the stimulus was delivered during diastole. (b) The effect of stimulation at moderate frequency, starting at the first arrow, is to produce progressive summation (7) leading eventually to tetanus (8). Jones, 1964 based on Yeager.

phase of contraction (systole), a phase of relaxation (diastole) and a phase of diastasis, the vessel pausing for a time in the relaxed condition. The slight presystolic notch is thought to reflect a passive expansion of the vessel associated with the build-up of pressure during the contraction of adjacent regions. The heart appears to remain excitable at all stages of the contraction cycle, the refractoriness which is so characteristic of the vertebrate heart being completely absent. Electrical stimulation in early systole leads to summation of the induced contraction, and in diastole or diastasis it produces an early contraction (Fig. 3.4(a)). In fact, it is possible to tetanize the heart of a cockroach, with normal rhythmicity completely abolished (Fig. 3.4(b)). The mechanical activity is associated with complex changes in electrical potential, comprising partially fused fast diphasic and slower monophasic waves.

The rhythmic contraction of the dorsal vessel appears to be myogenic in many insects, and fragments of vessel completely isolated from the body may continue to beat, though at a much slower rate than normal. Although the beat may be initiated inherently, it is subject to control by nervous and humoral influences. A pair of lateral nerves accompany the dorsal vessel, made up of segmental branches from the ventral ganglia and from the cardiac ganglia of the stomatogastric system, and the modifying influence of this innervation has been demonstrated by electrical stimulation of associated parts of the nervous system, which may cause arrest of heart beat, or acceleration of the beat frequency, depending on the rate of stimulation.

The action of a variety of pharmacologically active substances has been investigated with various insect heart preparations. Acetylcholine appears to have a stimulating effect on heart rate, an effect which can be blocked by atropine and curare, and potentiated by eserine, suggesting that cholinergic pacemakers may be involved. The effects of other drugs, like nicotine and adrenalin, on isolated heart preparations are rather variable and difficult to interpret in the present state of knowledge. There can, however, be little doubt that humoral control is of importance, in view of the isolation from the corpora cardiaca of substances capable of causing a marked acceleration of the beat frequency in semi-isolated preparations. The active principle has been partially purified, and appears to be a peptide; its action is apparently an indirect one, in that it causes the release from pericardial cells of a cardiac stimulator. The increased rate of heart beat which occurs after feeding in a number of insects is perhaps the outcome of activity in this chain of effectors.

4. Conclusion

This brief review of the insect circulatory system has focussed attention on the simpler and more regular aspects of its activity, and may therefore have left an impression of a fairly well organized system providing for the efficient transport of materials from one part of the insect's body to another. The reality of the situation is probably quite different, and it may be well to list a few of the more startling observations concerning the aberrancy of the physiology of circulation, to serve as a corrective against too superficial a view. It has, for instance, been shown that the heart can be removed completely from some insects without causing their death; and in many, the heart may stop for long periods without death or "evident dismay" on the part of the insect. Reversal of the direction of beat is commonly observed in many different species, and in most the rate of beat may vary between very wide limits. The flow of haemolymph in the open system of sinuses shows corresponding irregularities, sometimes moving in jerks synchronous with contractions of the dorsal vessel; often it is slow and discontinuous and in some regions of the body it may cease altogether for long periods. Experiments involving the injection of dyes or of radioactive material into the haemocoele have shown that the mixing of haemolymph is relatively inefficient, and it may take as long as 15 min for injected material to move from a superficial sinus to the dorsal vessel. Observations of this kind suggest that the analogy between the circulatory system of vertebrates and that of insects should not be pushed too far; although they may have many functions in common the parallel appears to break down in a number of cases, and it seems likely that this may be associated with corresponding divergences at the level of circulatory mechanism.

One of the most striking differences between the circulating medium of insects and of vertebrates is, of course, that in vertebrates it serves as a carrier of oxygen, while in insects, with their extensive tracheal system (see Chapter 7) the haemolymph would play a marginal role in oxygen transport. This difference has enormous implications in terms of functional organization; in vertebrates the demand for oxygen by every cell in the body is urgent and constant, and the system responsible for its transport must show a corresponding level of efficiency; where respiratory requirements are met by a tracheal system, the transport function of blood may be imagined to become much less demanding, since it will involve materials like carbohydrates which can be carried in quite concentrated solution, or materials like hormones, where fast supply is not a primary requirement. Under these circumstances one could imagine that the selection for improvements in the system as a transport system might become subordinate to selection for other aspects of function. Of these, the most important might be the function of haemolymph as a store, both of water and of dissolved materials of various kinds. The importance of this function is clearly indicated by the high concentration of soluble materials, and by the no less striking changes in haemolymph volume, which are very difficult to reconcile with the notion of an efficient transport system. It is common knowledge that it is possible to draw quite large quantities of haemolymph from many species of insect, provided they are in the appropriate physiological state, while at other times and in other species the insect may appear almost dry, and the quantity of haemolymph is clearly exceedingly small. Haemolymph volumes have been determined for a number of different species, and values ranging between 0.9% and 45.4% of the total body weight have been reported. Wide variations occur within a species depending on the stage of development and general physiological condition. The changes associated with ecdysis are particularly striking, as shown in Fig. 3.5. There can be little doubt that the marked rise in haemolymph volume which occurs at the time of moulting is associated with the function of haemolymph in providing for the hydrostatic distribution of pressure to ensure the proper expansion of the cuticle, a process which will be discussed further in Chapter 13. Such changes in blood volume have been shown to be in part attributable to a shift of fluids between haemocoele and intracellular reservoirs, and they are likely to become of particular importance during periods of

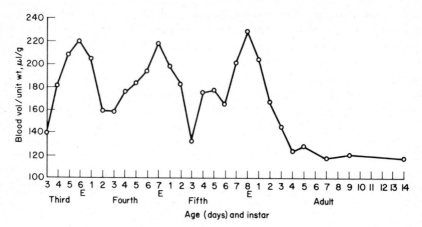

Fig. 3.5. Variations in the volume of blood per unit weight of insect during the development of the locust from the third instar. E denotes the time of ecdysis (Lee, 1961).

desiccation, when the haemolymph could act as a reserve of water for the tissues.

Once it is appreciated that functions of the haemolymph other than transport may be of paramount importance, the apparent inefficiency of the circulatory system in a class of animals which otherwise embodies so many elegant adaptations ceases to be an embarrassment, and certain general features of insect organization may then be seen as bound up with the problems that would arise when transport function is made subsidiary. The tendency for the different organ systems of insects to be extraordinarily diffuse, compared with their counterparts in higher animals, is an example. The tracheal system of necessity ramifies to all parts of the insect body, but the fat body, too, considered by some as in certain respects analogous to the liver of higher animals, is remarkably diffuse, and there are few parts of the insect's body which are far removed from cells of this tissue. The excretory organs, also, tend to be distributed widely, with tubules of the Malpighian system ramifying extensively in the abdomen, and extending into the thorax in many species. It seems not unlikely that such a broadcasting of the main organ systems may be related to the inefficiency of circulation, to the likelihood that stagnant pools may form in the haemocoele, and that such pools would have to rely on diffusion rather than on bulk flow for the purging of toxic products and for the replenishment of reserves; the proximity of an excretory tubule to such a toxic pool, or of a collection of fat body cells, would ensure that the path of diffusion remains a short one, and the process of diffusion therefore relatively efficient. The high concentration of many metabolites in the haemolymph is of obvious relevance in this context, since this would serve to maintain a steep diffusion gradient between haemolymph and site of metabolism, and the occurrence of a variety of enzymes in the haemolymph may also be pertinent, serving to transform substances diffusing into the haemolymph, and thus to maintain the diffusion gradient. It may even be that one of the basic insect characteristics, namely small size, should be seen in part as a reflection of circulatory inefficiency; for it is likely that only an animal of relatively small dimensions could rely as extensively as an insect appears to do on the process of liquid diffusion, for the satisfaction of many of its somatic requirements.

CHAPTER 4

FOOD STORAGE

The products of digestion, following their absorption from the alimentary canal, may be used directly to satisfy metabolic requirements, or they may be deposited in food stores from which they can be drawn in times of need. The cells of a variety of tissues (e.g. muscle cells and cells of the midgut) may be involved in food storage, and the haemolymph itself has been shown to constitute an important reserve of organic materials, but the main storage organ of insects is the fat body. The most important storage materials are glycogen and fat, and some indication of the extent of fat body development is given by the fact that, in some insects, the amount of fat may exceed 50% of the total dry weight of the animal, while glycogen contents in excess of 33% of dry weight have been reported.

The fat body consists of aggregates of cells forming lobes and sheets of tissue, which invest the internal organs of the body and constitute a conspicuous element of anatomy in the well-fed insect. The diffuse arrangement of adipose tissue is such as to facilitate exchange of material between it and the haemolymph which bathes it. The cells of which it is composed are known as trophocytes, and during periods of food intake they grow in size and become filled with droplets of fat and protein, and with particles of glycogen (see Fig. 4.1 (a), (b)). In the well-fed insect the droplets of fat tend to dominate the appearance of the fat body cells; mitochondria are distributed widely and parts of the cytoplasm show a well-developed endoplasmic reticulum bearing ribosomes, to give the characteristic "rough surface" appearance. The regions which are free of endoplasmic reticulum are packed with minute granules of glycogen.

During periods of food deprivation there is a gradual reduction in the amount of stored material; glycogen granules become less numerous and the fat droplets disappear. With prolonged starvation, degenerative changes set in, and the so-called albuminoid granules make their appearance. These are thought to represent remnants of cell organelles undergoing degenerative changes as a result of lysosome activity, the setting up of special regions of the cytoplasm to promote histolysis of intracellular material. There thus appears to be a general

56

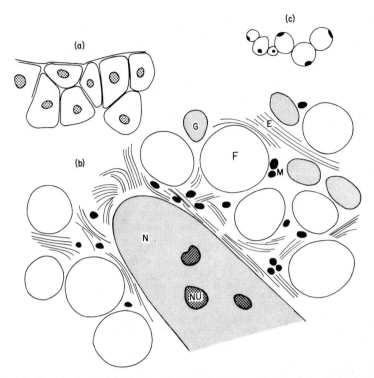

Fig. 4.1. The structure of fat body. (a) General arrangement of cells in the fat body. (b) Diagram of a cross-section of part of a fat body cell. N, nucleus; NU, nucleolus; F, fat droplet; G, glycogen-rich granule; M, mitochondria; E, endoplasmic reticulum (drawn from electron micrograph of Ishizaki, 1965). (c) A cluster of fat droplets from the fat body of an insect, each with its "cap" of lipase, as revealed by the 5-bromoindoxyl acetate method (drawn from photomicrograph of Wigglesworth, 1958).

erosion of the very fabric of the fat body, and the materials which are in this way released presumably serve to buttress other aspects of the insect's metabolism.

There can be little doubt that cyclical changes of a less extreme nature are a feature of the day-to-day existence of many insects. These have not been studied at the level of cellular ultrastructure, but they find reflection in periodic fluctuations in the quantity of reserve. These range from long-term changes, such as the gradual build-up and decline of fat reserves which take place during the life of a tsetse fly (Fig. 4.2(a)); though shorter cyclical changes in fat content, such as have been observed particularly in blood-sucking insects (Fig. 4.2(b)), but would be associated with intermittent feeding or diurnal fluctuations in feeding intensity in many other species; to the quite rapid depletion of glycogen and fat reserves which have been demonstrated for fruit flies and

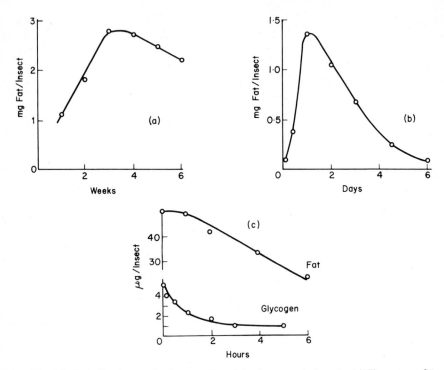

Fig. 4.2. Periodic changes in the quantity of food reserves in insects. (a) The average fat content of a population of tsetse flies in their natural environment, during successive weeks of adult life (from Jackson, 1946); the curve shows a general trend on which day-to-day fluctuations of individual flies (as shown in (b)) would be superimposed. (b) Fluctuations in the fat content of tsetse flies during the course of a single hunger cycle following a blood meal on Day 0 (from Bursell, 1963). (c) Changes in the fat and glycogen content of aphids during successive hours of flight (data from Cockbain, 1961).

aphids during flight (Fig. 4.2(c)); in these insects glycogen is used predominantly during early phases of flight, while fat becomes the principal substrate later.

Long-term changes in the quantity of stored food materials occur also during development, where insects may be deprived of the opportunity to feed for extended periods. In the mealworm beetle, and in the silkworm, a great deal of the glycogen laid down during late stages of larval life is utilized for pupal development; in the tsetse fly metamorphosis appears to be sustained almost entirely by fat reserves, which become greatly depleted in the course of pupal development.

Granted that cells of the fat body perform the function of storage organs for the main classes of food substance—carbohydrates, fats and proteins—the question arises whether the fat cells are themselves responsible for the synthesis of these storage materials from their constituent components—monosaccharides, fatty acids and amino acids—or whether the storage compounds are wholly or partially synthesized elsewhere and merely taken up for storage by the fat cells.

As far as carbohydrates are concerned, there is evidence that enzymes capable of mediating the synthesis of glycogen from glucose are present in the fat body and also that the fat body is closely associated with the metabolism of trehalose. Radioactive glucose has been found to be converted quickly to trehalose by fat body *in vitro*, while other tissues, such as muscle, blood and gut were found to be relatively inactive. The enzyme trehalase has also been demonstrated in fat body, but its activity there is rather slight compared with other tissues, and it would seem that carbohydrate is mobilized from the fat body as trehalose, and that it is transported in this form to the site of respiration.

The stored fats are present mainly as neutral triglycerides, although free fatty acids may make some contribution to the total. Evidence has recently become available for the existence of an enzyme system in insect fat body capable of synthesizing free fatty acids. Lipases, responsible for the hydrolysis of glycerides to their constituent fatty acids and glycerol have also been shown to be highly active in fat body homogenates, and it has been possible to demonstrate the presence of a small "cap" of lipase associated with each intracellular droplet of fat (see Fig. 4.1(c)). It seems likely, therefore, that fat is not only stored, but also synthesized by the fat body, and that its mobilization from the fat body involves the liberation of free fatty acids from their neutral fats, such fatty acids being transported to respiring tissues for further breakdown.

The situation as far as protein is concerned is not altogether clear. The ability of fat body cells to synthesize proteins, associated with the presence of a "rough" endoplasmic reticulum, has been demonstrated convincingly in studies involving haemolymph proteins. Various amino acids have been shown to become incorporated into proteins which are subsequently released from the fat body into the haemolymph. But the precise nature of the proteins which can be shown by histochemical techniques to be stored in intracellular droplets of the fat body cells remains obscure, as does the mechanism of their mobilization.

In addition to the storage of food materials, certain cells of the fat body may serve as stores of uric acid. The extent to which this occurs differs from species to species, but the phenomenon would appear to be particularly common among the Orthoptera. In certain cockroaches, placed on a diet rich in nitrogen, the fat body may become greatly enlarged through the distension of its cells with white deposits of uric acid. Homogenates of fat body have been shown to be active in the synthesis of uric acid, but the significance of its accumulation in fat body cells is not yet clear. It seems likely that it represents a process of storage excretion, and it will be discussed further in the section dealing with nitrogenous waste (Chapter 6). On the other hand, some authors have suggested that it may represent a storage of nitrogen, which could be mobilized for synthesis of nitrogen-containing compounds during periods of nitrogen deprivation. This is an attractive idea, but as yet there is no evidence that uric acid is in fact broken down in such a way as to make its nitrogen available for such a purpose.

Apart from their function as a store of the three main types of food material, the fat body has been shown to perform an extremely important function in intermediary metabolism. Its potentiality in this respect has been under intensive investigation during the past few decades, and an impressive spectrum of functional activity has been demonstrated, based on the presence of many different kinds of enzymes including oxidases, dehydrogenases, transaminases, esterases and phosphatases. The special metabolism of purines and certain eye pigments appears to be the particular province of the fat body.

In view of its importance in intermediary metabolism, of its role as a store of food reserves and of the part that it plays in detoxication, it is not surprising that the fat body has been regarded as the insect equivalent of the vertebrate liver.

CHAPTER 5

OSMOREGULATION

The composition of insect haemolymph has been discussed in an earlier chapter, and it was noted that the osmotic concentration differed from species to species, and that different patterns of ionic composition characterized different groups of insects. In some, sodium predominates among the cations, in others potassium, while in the Phasmidae, magnesium is present in higher concentration than the other cations; and similar, though less extreme, variations occur among anions. It would seem that for a given species of insect, at a given stage in its life history, there is a general level of total osmotic concentration, and a general pattern of proportionate composition, and that these serve as the basis for normal physiological function. The question then arises as to the mechanism by which the haemolymph composition is maintained at this appropriate level.

1. The Nature of the Problem

For an insect, as for any other terrestrial animal, there are two main factors which would tend to upset the osmotic and ionic equilibria:

(a) losses of water by transpiration, and gain of water by drinking would tend to concentrate or dilute the body fluids, and lead to a change in osmotic pressure without affecting proportionate composition; and

(b) intake of inorganic salts with the food would lead to an upset of osmotic balance if the osmotic concentration of the food is different from that of the haemolymph, and to ionic imbalance if the ionic composition of the food is different from that of the haemolymph.

It is possible, therefore, to distinguish two different aspects of regulation: the ability to regulate total osmotic pressure, and the ability to regulate ionic composition. In terms of mechanism, however, both depend on the capacity for active transport of solutes, and perhaps of water, on the part of certain specialized epithelia, and the distinction will not be maintained in the discussion that follows.

The problem of osmoregulation in terrestrial insects is thus essentially that of getting rid of such water and salts as may be ingested in excess of requirements arising as a result of losses of water by transpiration and excretion, and of salts by excretion. As far as the salts are concerned, one way, theoretically, of maintaining regulation would be to limit the absorption of salts from the alimentary canal—to take up mainly the necessary organic constituents and to allow most of the inorganic material to pass through as faecal matter. In fact, inorganic ions are quickly absorbed from the alimentary canal, ingested salts therefore mix with the general pool of inorganic material in the haemolymph, and the problem arises of sequestering them in some way from this pool. This becomes a particularly difficult task where the proportion of different ions in the food differs substantially from that which characterizes the haemolymph, and this is often the case, as shown by the values listed in Table 5.1. A primary

TABLE 5.1

The cation composition of the haemolymph of some terrestrial insects compared with that of their food

	Concentration m.eq/l or m.eq/kg wet wt.			
	Na^+	K^+	Ca^{++}	Mg^{++}
Gastrophilus intestinalis (larva)	175	12	6	32
Whole horse blood	85	31	2	3
Dixippus morosus (adult)	9	28	16	142
Privet leaves	46	152	825	40
Leptinotarsa decemmineata (adult)	4	65	48	189
Potato leaves	trace	145	129	86

Data for *Gastrophilus* from Levenbook (1950); otherwise Duchâteau *et al.* (1953).

requirement for sequestration would be the production of a urine, which could serve as a vehicle for the removal of inorganic constituents, and provision would then have to be made for the resorption from this urine of water and of such constituents as would be required to maintain the normal composition of the haemolymph. In insects the Malpighian tubules serve in the formation of urine, while the rectal glands of the hindgut are active in the resorption of wanted constituents, and it will be useful to defer consideration of physiological mechanisms until an account has been given of the structure of these main elements of the excretory system.

2. Structure of the Excretory System

A generalized diagram of the excretory system of an insect is shown in Fig. 5.1(a). The excretory, or Malpighian, tubules lie free in the haemolymph of the abdomen, ramifying among the other abdominal organs. They are closed distally, and open proximally to the alimentary canal at the junction of midgut and hindgut. The number of tubules varies between species, from two to several hundred, and there may be some histological differentiation between tubules, or between different regions of a single tubule. The functional significance of such differentiation has not yet been fully elucidated, and it will not be described in detail. The tubules are lined with a single layer of epithelial cells supported by a conspicuous basement membrane, and associated with each tubule is a spiral band of muscle cells, whose contractions cause writhing movements of the tubules, which may assist in the bulk flow of urine; they are often accompanied along their length by tracheoles.

The epithelial cells of the Malpighian tubules are characterized by a "brush border", and electron micrographs show that the inner surface is extensively folded to produce a layer of tightly packed microvilli, many of which contain elongated mitochondria (see Fig. 5.1(b)). The basal plasmalemma is also extensively folded, and these invaginations, too, are associated with a high density of mitochondria. The submicroscopic structure thus conforms to the pattern of cells which are engaged in intense secretory activity.

The contents of the Malpighian tubules enter the alimentary canal to mix with end-products of digestion from the midgut, and pass backwards through the hindgut to the rectum. The epithelial cells of the hindgut and rectum are covered by a distinct cuticular lining, and are invested by incomplete layers of muscle, whose contractions cause deformations of the rectal chamber, and thus efficient mixing of the rectal contents. A conspicuous feature of the rectal anatomy of many insects are the so-called rectal glands, whose precise form varies from species to species. In some they are represented by longitudinal bands of well-developed glandular cells (Fig. 5.1(c)); in others they may form distinct papillae, projecting into the rectal cavity (Fig. 5.1(a)); in all species, even those that show no marked differentiation between glandular patches and the general rectal epithelium, the organs are richly supplied with tracheoles.

3. The Production of Urine

The primary requisite for ionic regulation in terrestrial animals is the formation of urine. In view of the low-pressure, open and often erratic nature of the blood circulation in insects (see Chapter 3), the possibility of using some form of filtration system as a basis for urine formation, as in the vertebrate kidney, is clearly excluded, and recourse must be had to a quite different

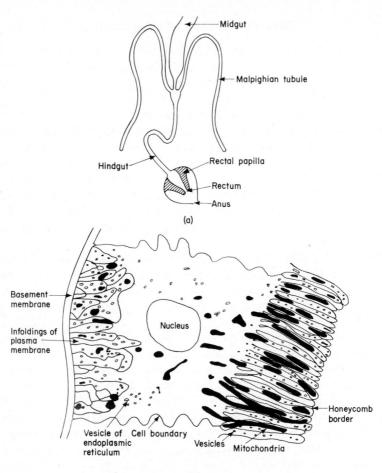

(a)

(b)

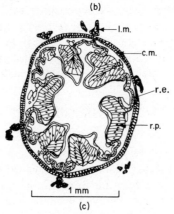

(c)

principle. It has in fact been demonstrated that the urine of insects is formed by an active secretion of inorganic ions, potassium in particular. This was brilliantly established by the work of Ramsay, who developed techniques for the collection and analysis of minute quantities of fluid from different parts of the excretory system of a number of different species of insect, and his results have been fully confirmed by later workers. Figure 5.2(i) shows one of the simple yet delicate systems employed by Ramsay in his original investigation. Segments of Malpighian tubule were isolated in a drop of haemolymph under liquid paraffin, with an adequate supply of oxygen assured. The severed end of the tubule was drawn out of the drop of haemolymph by means of a fine silken thread, and a cut was made in the tubule close to the silk ligature, allowing the urine produced to escape as a droplet round the ligature, from which it could be collected for analysis. It was found to be approximately isosmotic with the haemolymph, but its potassium concentration was very much greater (see Table 5.2). This inequality in the distribution of potassium between haemolymph and tubule could not be accounted for on the basis of an electrical gradient across the tubule wall; in *Dixippus*, for example, the tubule content is positive with respect to haemolymph, which would tend to oppose an inward transfer of potassium; and even in insects where the tubule interior carries a negative charge (as in the locust and in *Rhodnius*), the magnitude of the potential difference is not commensurate with the difference in potassium concentration. It would seem, therefore, that potassium is transported from the haemolymph across the tubule epithelium and into the lumen against an electrochemical gradient, and that the process must therefore involve some active mechanism. The potassium ions are probably accompanied by chloride to preserve the electrical equilibrium, so that what is effectively involved is a transport of neutral salt across the tubule epithelium. The effect of such an active transport would be to increase the osmotic pressure of the tubule fluid, yet its osmotic concentration does not differ greatly from that of haemolymph (see Table 5.2). This shows that the active uptake of salts is associated with an isosmotic uptake of water; this, in turn, would increase the hydrostatic pressure inside the tubule and cause a bulk flow of fluid along it, in other words, a flow of urine.

Further work on isolated Malpighian tubules has amply confirmed this concept of urine formation. The special role of potassium has been substantiated by investigations of the effect of ionic composition of the medium bathing the tubules on the rate of urine production, and on the composition of urine. As

---

Fig. 5.1. Features of the excretory system of insects. (a) Diagram of the interrelationship between the alimentary canal and the excretory system in an insect. (b) Section through a cell of the distal region of the Malpighian tubule of *Rhodnius* (Stobbart and Shaw, 1964, from Wigglesworth and Salpeter). (c) Diagram of a cross-section of the locust rectum. c.m., circular muscle; l.m., longitudinal muscle; r.e., reduced epithelium between rectal pads; r.p., rectal pad (Phillips, 1964).

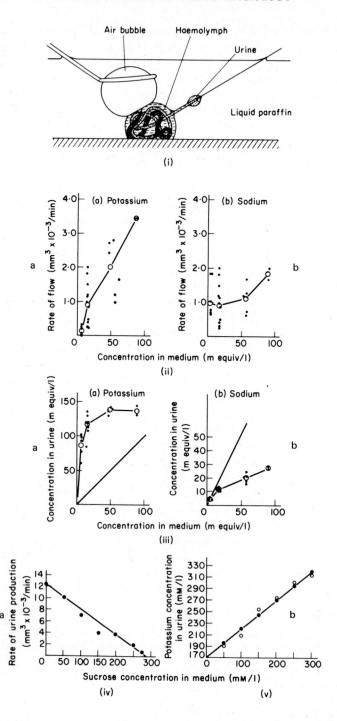

shown in Fig. 5.2(ii), the rate of urine flow is closely dependent on the potassium concentration of the medium, while it is relatively unaffected by changes in the concentration of sodium; and Fig. 5.2(iii) shows that while the potassium concentration of the urine increases sharply with an increase in haemolymph potassium, the concentration of sodium is relatively unaffected by changes in the sodium content of the medium. Recently it has been shown that urine production is also sensitive to the osmotic pressure of the medium (Fig. 5.2(iv)), and that maintenance of urine flow in media of high osmotic pressure is associated with a corresponding increase in the potassium concentration of the urine (Fig. 5.2(v)).

The precise mechanism by which the transport of potassium and water across the tubule epithelium is achieved has not yet been fully elucidated, but it is considered that the complex infoldings of both basal and apical membranes (see Fig. 5.1(b)) are of significance in enabling the setting up of standing osmotic gradients. Ions which have been actively transported into such submicroscopic reservoirs would be prevented from diffusing away by structural barriers, thus enabling water to follow along the osmotic gradients.

With regard to ions other than potassium, and to organic molecules such as amino acids and carbohydrates, consideration of the corresponding electro-chemical gradients suggest that these enter the tubules by passive diffusion from the haemolymph. The most important exception is uric acid, which appears to be actively secreted into the tubule against a steep concentration gradient (see Table 5.2), probably as the soluble sodium or potassium salt.

4. Rectal Resorption

With the formation of a urine containing a variety of inorganic ions which have entered the Malpighian fluid by secretion and by passive diffusion, the indispensable basis for ionic regulation has been secured, but without further manipulation of the urine so produced the sole ionic effect of the process would be a progressive potassium impoverishment. A mechanism enabling selective

---

Fig. 5.2. Aspects of the physiology of excretion in insects. (i) Isolated Malpighian tubule preparation, see text for explanation (Ramsay, 1954). (ii) (a) The rate of urine flow in the stick insect as a function of the potassium concentration of the medium (Na^+ at 16-17 m.eq/litre). (b) The rate of urine flow as a function of sodium concentration of the medium (K^+ at 15-16 m.eq/litre) (Ramsay, 1955). (iii) (a) Potassium concentration in the urine of the stick insect as a function of potassium concentration in the medium (Na^+ at 16-17 m.eq/litre). (b) Sodium concentration in the urine as a function of the sodium concentration in the medium (K^+ at 15-16 m.eq/litre): the straight lines indicate equal concentration in urine and medium (Ramsay, 1955). (iv) The effect of sucrose concentration in the medium on the rate of urine flow in the blowfly (Berridge, 1968). (v) The effect of sucrose concentration in the medium on the potassium concentration in the urine of the blowfly (Berridge, 1968).

TABLE 5.2

Ionic composition of the haemolymph, urine and rectal contents of the stick insect and the locust

Source of fluid	Osmotic pressure NaCl eq	Uric acid mg/100 ml	Concentration m.eq/l						Potential difference mV
			Na^+	K^+	Cl^-	Ca^{++}	Mg^{++}	$PO_4^{\equiv}$	
Dixippus morosus									
Haemolymph*	171	5	11	18	57	7	108	39 ⎫	+ 21
Urine	171	43	5	145	65	2	18	51 ⎭	
Rectal contents	390	–	18	327	–	–	–	–	
Schistocerca gregaria									
Haemolymph	214	–	108	11	115	–	–	–	
Urine	226	–	20	139	93	–	–	–	
Rectal contents (water fed)	433	–	1	22	5	–	–	–	
Ditto (saline fed)	989	–	405	241	659	–	–	–	

Data for *Dixippus* from Ramsay (1953 and 1955); data for *Schistocerca* from Phillips (1964).

* Values are for the "serum" separated by centrifugation from the clot formed by heating the haemolymph to 100° for 5 min.

resorption of urine components needs to be superimposed in order to achieve ionic regulation, and such a mechanism is provided by the activity of the rectal glands, or the rectal epithelium in insects which lack distinct glands.

The important part which the rectal epithelium plays in ionic regulation is witnessed by the changes in composition which occur after the urine has entered the rectum (see Table 5.2). The total osmotic concentration shows a substantial increase, but the extent to which this is reflected in the concentration of inorganic ions depends on ionic balance. In starved animals which have access to water, the need to conserve inorganic salts would be paramount, and the concentration of sodium, potassium and chloride in the rectal fluid is reduced to extremely low values. The high osmotic pressure of the rectal contents of the locust in these circumstances must then be attributed to organic solutes that in the absence of specific information, may be classed as waste products. In feeding insects, or in saline-fed insects, on the other hand, there would be a need to eliminate excess ions, and their concentration in the rectal fluid remains at a correspondingly high level, accounting for a substantial proportion of the total osmotic pressure.

These changes in composition suggest that the rectum is active in the resorption both of water and of inorganic salts, and that it is here that the composition of the urine is accurately adjusted to make its excretion effective in ionic regulation. The mechanisms involved in rectal function have been investigated by Phillips, whose work will form the basis of the account that follows. The uptake of water and salts by rectal glands of the locust was determined following injection of appropriate solutions into the rectum, separated from the rest of the alimentary canal by ligatures at the junction of midgut and hindgut. Radioactively labelled serum albumin, a substance which is not taken up by the rectal epithelium and cannot diffuse across it, was incorporated in the injection fluids to provide a convenient check on changes in liquid volume associated with the uptake of water.

a. Resorption of Water from the Rectum

When a solution of a carbohydrate which is not absorbed, such as xylose or trehalose, is injected into the rectum of a fasting locust, water is resorbed from the solution even though its osmotic pressure may be substantially above that of the haemolymph. This absorption of water against a concentration gradient is not associated with a significant net uptake of inorganic salts, which raises the possibility that an active transport of water may be involved. It is possible, however, that at the submicroscopic level the movement of water may be an isosmotic one, associated, perhaps, with a cycling of potassium ions across subcellular membranes. Whatever the precise mechanism, the uptake has been shown to be limited by the osmotic gradient, as illustrated in Fig. 5.3(a). With dilute solutions, where the osmotic gradient between rectal contents and haemolymph would favour an uptake of water from the rectum, absorption takes place at the rate of about 40 μl/hr. In the absence of a favourable osmotic gradient the rate of uptake is substantially reduced, but it is not until the osmotic concentration of the rectal contents greatly exceeds that of the haemolymph that net uptake ceases altogether; and what is of particular interest in relation to the general problem of osmoregulation and water balance is, that the level to which osmotic concentration of the rectal contents can be raised by absorption of water varies according to the water requirements of the insect. In starved individuals with access to fresh water, the resorption of water ceases when the osmotic pressure reaches a value equivalent to a freezing point depression of $1.5°$, while in saline-fed insects freezing point depressions approaching $3.0°$ have been recorded.

b. Resorption of Solutes from the Rectum

To study the absorption of inorganic ions from the ligated rectum of the locust, injection fluids made hyperosmotic with xylose were used, in order to avoid complications which might arise from a simultaneous absorption of water. In this way it was shown that an uptake of sodium, potassium and chloride can occur against a strong electrochemical gradient, and that in water-fed insects the end result is the all but complete removal of these ions from the rectum (see Table 5.2). The rate of absorption depends upon rectal concentration and potassium is taken up very much faster than sodium (Fig. 5.3(b)). The rate of chloride uptake is higher when potassium predominates than when sodium predominates, suggesting that its uptake is linked with that of the cation. Again, the process appears to be closely geared to requirements in the sense that for saline-fed insects the rate of potassium uptake is greatly reduced at high ion concentrations (Fig. 5.3(b)), thus providing a mechanism by which a copious excretion of excess ions may be effected.

The uptake of substances other than water and inorganic ions from the rectal fluid has not been the subject of detailed study, but since a variety of solutes are known to enter the fluid of the Malpighian tubules, yet fail to make their appearance in the insects' excreta (see Chapter 6), it must be assumed that a process of resorption occurs in the rectum. Further studies will be needed to establish whether such absorption takes place by an active process against an electrochemical gradient, or whether it is a simple consequence of an increase in rectal concentration associated with the uptake of water.

One substance which is not resorbed by the rectal glands is uric acid, which enters the rectum as the soluble potassium salt in alkaline solution. In the rectum the pH of the urine is reduced from about 7.0 to between 4.0 and 6.0, possibly as the result of an active secretion of hydrogen ions by the rectal epithelium. This causes the formation of highly insoluble uric acid from its more soluble salt and, coupled with the resorption of water from the rectum, results in

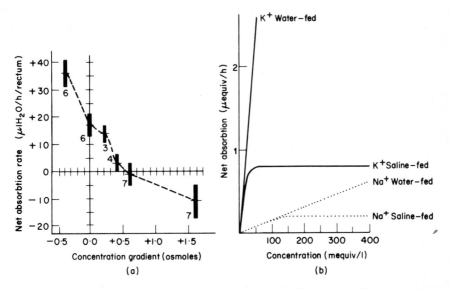

Fig. 5.3. Aspects of rectal absorption in the locust. (a) The relationship between the initial osmotic gradient and the rate of water movement across the rectal wall. A positive sign indicates absorption from the rectum (ordinate), or rectal fluid more concentrated than the haemolymph (abscissa); bars indicate the extent of standard deviations, and subscript numbers the number of observations (Phillips, 1964a). (b) The relationship between rectal fluid concentration and the net rate of absorption of potassium and sodium from the ligated rectum of the locust (redrawn from Phillips, 1964b).

a massive precipitation of uric acid. In view of the limitations set to the absorption of water by the osmotic gradient, the occurrence of uric acid as the main excretory product in insects is clearly of the greatest significance. The maximum osmotic effect which it can exert will be governed by its solubility in an acid medium, which is no more than 0.0004 osmoles/litre (6 mg/100 ml). Provided that other solutes are removed, water can therefore continue to be resorbed from the rectal contents without any increase in its osmotic concentration.

5. Regulatory Aspects

An attempt has been made to bring together the results described in previous sections in the form of a summary diagram, to serve as the basis for a discussion of regulatory aspects of excretion. Figure 5.4 shows the secretory cycling of potassium and water from haemolymph to Malpighian tubules and from rectum back to haemolymph, which forms the basis of urine formation and of osmoregulation. Also shown is the active secretion of uric acid, and the passive diffusion of other solutes, into the Malpighian tubule, followed by the

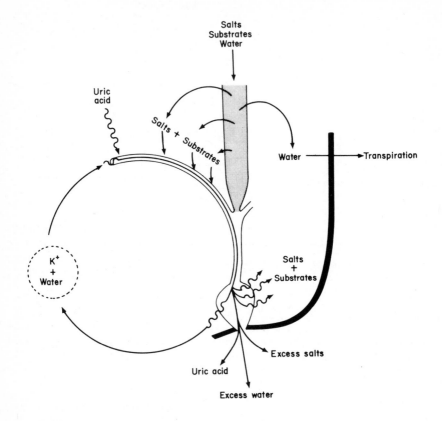

~~~▶ Active secretion

Fig. 5.4. A diagrammatic summary of processes involved in the osmoregulation of insects; for further explanation see text.

elimination of uric acid from the rectum, and the recovery of salts and substrates, whether by active or isosmotic resorption. The different components of this system have been shown to be appropriately influenced by the water and salt requirements of the insect for it to serve as an effective basis of regulation. For instance, under conditions of desiccation, an increase in haemolymph osmotic pressure would cause a reduction in urine flow (see Fig. 5.2(iv)), and an increase in the absorption of water from the rectum (p. 70); the net effect would thus be to minimize excretory losses of water at a time when transpiratory losses are high. If the desiccating conditions involve an uptake of food with a high salt content, there would in addition be a decrease in the absorption of salts from the rectum (see Fig. 5.3(b)), and osmotic equilibrium would tend to be maintained in the face of declining water reserves by a correspondingly copious elimination of salts.

In insects under conditions tending to hydration, as in saturated atmospheres with access to fresh water but not to food, the danger would lie in the possibility of excessive dilution of body fluids. This would be counteracted by the increased urine flow and decreased rectal absorption of water associated with low haemolymph osmotic pressure; and coupled with an increase in the absorption of salts from the rectum, the result would be an elimination of water with limited loss of inorganic ions.

Some of these regulatory effects (e.g. increase in osmotic pressure leading to a decrease in rate of urine formation) may reasonably be attributed directly to changes in the magnitude of electrochemical gradients across the secretory epithelia. But the decrease in the absorption of salts, and increase in the absorption of water, which characterizes saline-fed animals must clearly be based on some indirect effect on the activity of the secretory cells, and it seems that effects of this kind may be mediated by blood-borne factors of the neuroendocrine system. One of the best documented examples of the hormonal control of excretion is furnished by the work of Maddrell on the blood-sucking bug *Rhodnius*, in which intense diuresis occurs soon after feeding. This coincides with the appearance in the haemolymph of a very active diuretic hormone, which appears to be released from a series of swollen nerve fibres associated with the abdominal nerves immediately behind the metathoracic ganglion.

While the mechanisms discussed above would provide a basis for the regulation of total osmotic pressure, the phenomenon of ionic regulation has still to be accounted for. This would involve a filtering off from the general haemolymph pool of ions which are taken in with the food, and would pose special problems where the proportionate composition of the food differs substantially from that of the haemolymph, as is often the case. The mechanisms by which such differential elimination of haemolymph ions is achieved have not yet been studied in sufficient detail to provide a firm basis of interpretation, but it seems that discrimination would occur at two sites:

(a) with an increase in the haemolymph concentration of a particular ion the rate at which it would diffuse into the Malpighian tubule, and hence the rate at which it would pass into the rectum, would increase; and

(b) the resorption rate of the ion from the rectum would be decreased, and become independent of rectal fluid concentration (ref. Fig. 5.3(b)) as a result of the increase in haemolymph concentration.

In other words, the rate at which the ion passes into the rectum increases, while the rate at which it is absorbed from the rectum decreases, so that the net result would be a selective elimination of the particular ion.

A great deal of effort has been devoted to the satisfactory elucidation of osmoregulatory mechanisms during recent years, but there has as yet been little work to determine the extent to which these mechanisms are effective in ensuring a regulation of ionic and osmotic concentration of body fluids in the

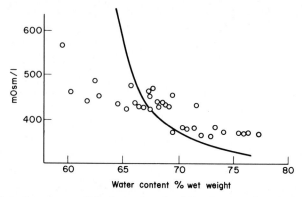

Fig. 5.5. Osmotic pressure of the haemolymph of the desert cockroach plotted against the insect's water content. The line represents the expected relationship in the absence of regulation (redrawn from Edney, 1966).

normal terrestrial insect. Changes in the haemolymph osmotic pressure of a desert cockroach during cycles of dehydration and rehydration are shown in Fig. 5.5, for comparison with the effects which would have been observed in the absence of regulation. The results show quite clearly that substantial changes in water content are reflected in relatively small changes in haemolymph osmotic pressure, indicating that regulatory processes are at work, and it is likely that some of the mechanisms discussed in this chapter may be involved; for instance an increase in the uptake of water, and a decrease in the resorption of salts, from the rectum during dehydration. But under the conditions of experiments of this type, with insects denied access to food, it is possible that the main factor involved in regulation may be an interchange of water and salts between haemolymph and tissue fluids. For a fuller elucidation of the mechanisms involved under natural conditions, and of their relative contribution to the total effect, determinations of haemolymph volume and of the water and salt content of the faeces would have to be included, and the experiment should preferably be carried out under conditions of controlled access to food and water, and under different conditions of desiccation stress.

CHAPTER 6

NITROGENOUS WASTE PRODUCTS

The disposal of surplus nitrogen is a problem that confronts most animals, because the intake of nitrogenous material is usually greatly in excess of the demand. It is only during periods of rapid growth that the reverse is likely to be true, and even then it will usually be the availability of some particular component which is limiting (see Chapter 2), so that a substantial amount of nitrogenous waste will still require to be excreted.

The problem of nitrogen disposal is an acute one, because although nitrogen itself is not a toxic substance, the form in which it generally appears in metabolism is, for the primary nitrogenous end-product both of amino acid and of purine metabolism is ammonia (see Chapter 1). The precise basis of ammonia toxicity has not yet been established, but it appears to be largely independent of the changes in pH which would be associated with the formation of ammonium hydroxide in aqueous media.

Certain aspects of the formation and disposal of nitrogenous waste products have been dealt with in earlier chapters. The metabolic pathways involved in the detoxication of ammonia, by incorporation of the nitrogen in uric acid, have been described in Chapter 1; and the mechanism by which uric acid, as the primary nitrogenous end-product, accumulates in the rectum for ultimate disposal by defaecation has been discussed in Chapter 5. What remains to be dealt with here are the quantitative aspects of the problem, the proportion of different nitrogenous waste products excreted by different insects, and methods of disposal of such waste products other than by normal excretion.

1. Excretion of Waste Products

It has long been recognized that uric acid is the main vehicle of nitrogenous excretion in terrestrial insects, and the class as a whole has generally been regarded as a satisfactory example of the so-called "uricotelic" mode of life, with uricotelism seen as a specific adaptation to the terrestrial habitat. It is argued that most aquatic animals are in a position to permit a relatively free exchange of small and readily diffusible molecules, like ammonia, between the internal and the external environment, either across the general body surface or across special respiratory epithelia. Under these circumstances they would be able to dispose of excess nitrogen by simple diffusion of ammonia across permeable membranes, at a rate which would be sufficient to prevent the build-up of ammonia in the body fluids to toxic levels. One of the primary requirements for the invasion of a terrestrial habitat was a drastic reduction in general permeability to water, which appears to have been accompanied by a corresponding reduction in the permeability to other molecules, like oxygen and ammonia. That it is the permeability of the integument which is of importance in this context, rather than simply the exchange of an aqueous for an aerial medium, is well illustrated by the terrestrial isopods, which have retained a permeable cuticle and continue to excrete the bulk of their nitrogenous waste as ammonia, by simple diffusion across the integument. In most other terrestrial groups, where ready diffusion of ammonia into the atmosphere is precluded, the need arose to develop some mechanism of detoxication, and this was met by the formation of urea in ureotelic groups and uric acid in uricotelic groups. Both of these substances, being non-volatile, would require an output of urine for their disposal, so that it is at this stage that the osmoregulatory system comes to take on a secondary role as an excretory system. Urea would serve as a satisfactory waste product in animals which had sufficiently large water reserves, or sufficiently ready access to water, to permit the loss of substantial quantities of water as hypotonic, or slightly hypertonic urine. In many terrestrial insects, however, access to water may be unreliable and infrequent, and because of their small size water reserves are minimal. For such insects an advantage would be gained by reducing losses of water by excretion, and this could not readily be done with urea as the main nitrogenous waste product, because of the high solubility of this substance. Withdrawal of water from the urine would be militated against by a progressive increase in the osmotic pressure exerted by such a soluble excretory product, and a low limit would be set to the hypertonicity of urine by the absorptive powers of the rectal epithelium (see Chapter 5). Under these circumstances the insolubility of uric acid would clearly give it a considerable advantage as an excretory product, because the osmotic pressure of a saturated solution of uric acid is well below the absorptive capacity of rectal epithelium, so that, in the absence of other soluble material, an all but complete withdrawal of water from the excretory product should be possible. The virtually dry excreta produced by many species of insect attests sufficiently to the importance of uric acid as a principal excretory product.

It should not be forgotten, however, that the insect pays a heavy price for the benefit which it derives in terms of water balance. The synthesis of a uric acid molecule involves the expenditure of substantial amounts of energy (see Chapter 1), and the four atoms of nitrogen which it contains are associated with five atoms of carbon, which might otherwise have featured to advantage in the context of synthetic and degradation metabolism. The nature of the bargain struck is well illustrated in blood-sucking insects, such as the tsetse fly, where the type of food imposes a particularly high nitrogen load. It can be calculated that for every 100 mg dry wt of blood ingested, no less than 47 mg have to be excreted to ensure the disposal of surplus nitrogen; and losses of energy associated with the manipulation of the blood meal (digestion, absorption, uric acid synthesis, excretion etc.) reduce the net gain to the insect to something like 50 mg of respirable material. If the insect had been able to dispose of the nitrogen by simple diffusion as ammonia, the corresponding value would be in the region of 85 mg.

The necessity to produce uric acid thus constitutes a serious disadvantage in terms of metabolism, but one that is outweighed by its advantage in the context of water balance, and uric acid has in fact been shown to constitute the main nitrogenous end-product in terrestrial insects. Within the limits of this broad generalization a situation of considerable complexity has been shown to exist, whose details unfortunately remain obscure for lack of sufficiently extensive information. The aim of investigators in the field of excretory metabolism has seldom been the evaluation of total nitrogen balance; many have been concerned simply to demonstrate the presence or absence of a specific material in the excreta, and even where information is available concerning a range of nitrogenous materials, this has rarely been coupled with determinations of total nitrogen, so that unequivocal assessment of the precise proportion which a given material, such as uric acid, constitutes of the total nitrogen cannot be made. For purposes of a general comparison between the few species for which reasonably extensive information is available, the nitrogen content of each substance has been expressed as a percentage of the total nitrogen content of the substances assayed. These will usually include the most important, though the possibility

TABLE 6.1

The percentage distribution of nitrogen among different excretory products in adult terrestrial insects

| Order | Species | % of total nitrogen | | | | |
		Uric acid + primary degradation products	Amino acids	Urea	Ammonia	Ref.
Orthoptera	*Melanoplus bivittatus*	55	11	4	29	1
Heteroptera	*Dysdercus fasciatus*	61	15	12	–	2
	Rhodnius prolixus	97	trace	3	–	3 + 4
Diptera	*Aedes aegypti*	66	7	15	12	5
	Glossina morsitans	82	15	trace	–	6

1. Brown, 1937
2. Berridge, 1965
3. Wigglesworth, 1931

4. Harrington, 1961
5. Irrevere and Terzian, 1959
6. Bursell, 1964

cannot be excluded that unidentified substances, or substances not quantitatively determined, may exceed the listed materials in importance, and Table 6.1 is presented with this qualification.

a. Uric Acid and its Primary Degradation Products

The nitrogenous excretion of insects has been the subject of fairly intensive investigation during recent years; in most of the species investigated, which include representatives of most of the orders, the bulk of excretory nitrogen appears in the form of uric acid, and in some virtually all of the excretory nitrogen is in this form. Even in species where uric acid is not the predominant excretory product it usually accounts for a substantial proportion of excretory nitrogen; and it is absent from the excreta only in a very few cases, where the use of other excretory end-products may be seen as a special adaptation to a particular mode of feeding. In certain plant-sucking insects, such as the cottonstainer, for instance, the need to dispose of large quantities of inorganic ions from the diet appears to have involved a shift from uric acid to allantoin as the main excretory product, and uric acid is completely absent from the excreta.

Substantial quantities of the primary degradation products of uric acid (allantoin and allantoic acid (see Chapter 1)) occur in the excreta of insects from most of the orders. In some, allantoin may be the predominant material, with uric acid and allantoic acid present in variable proportions ranging from zero to 30% or more of the total excretory nitrogen. In others, especially among the Lepidoptera, allantoic acid predominates, sometimes with allantoin and sometimes with uric acid as co-dominant. It has not been possible to establish any convincing correlation between the quantitative importance of one or other of these substances and other aspects of biology or of taxonomy. A given pattern of distribution of nitrogen among the three substances may occur among members of primitive or advanced orders, among insects with widely different modes of feeding, and indifferently in early or late developmental stages. The predominance of a particular end-product cannot, therefore, be regarded as a species characteristic, but becomes descriptive simply of a particular moment in the life history of a particular insect, and for this reason there seems at present little point in distinguishing too rigidly between the different patterns of excretion. Effort should be directed, however, towards an elucidation of their biological significance; the fact that the enzymes concerned with the primary degradation of the purine ring, uricase and allantoinase, appear to be particularly active in extracts from the Malpighian tubules and midgut, both of which have been shown to be associated with the elimination of excretory products, suggests that degradation of the purine ring may in some way be involved with the transfer of material across the secretory epithelium. But until this and other possibilities have been explored, it would seem preferable to regard the occurrence of high proportions of allantoin and/or allantoic acid as minor variations on a basic uricotelic theme, especially since these substances differ relatively little from uric acid in terms of the properties which are chiefly relevant in the context of excretory metabolism, namely nitrogen content and solubility. It would, on this view, be legitimate to retain the notion of terrestrial insects as predominantly uricotelic, defining a uricotelic animal as one that uses uric acid, allantoin or allantoic acid, or some mixture of the three, as the main excretory material.

b. Urea

Urea is present in the excreta of most insects which have been investigated, and it makes up a substantial proportion of total nitrogen in some (see Table 6.1), but the role of this material in nitrogenous excretion is poorly understood. Its presence cannot be accounted for on the basis of purine degradation, because no insect has been shown to possess the full complement of enzymes required to degrade uric acid to this stage. On the other hand, there is little evidence for the existence of an ornithine cycle in insects. The presence of arginase has been demonstrated in several species, catalysing the hydrolysis of arginine to urea and ornithine, and arginine itself is of importance as a phosphagen, a store of high energy phosphate, in insects. It is possible, therefore, that excretory urea may arise by the hydrolysis of arginine, but if so, the biological significance of the reaction remains to be elucidated.

c. Ammonia

Ammonia has been identified in the excreta of most insects in which it has been carefully looked for, and in some it may account for a substantial proportion of total nitrogen (see Table 6.1). Whether it should be regarded simply as a fraction of the total ammonia arising in the course of deaminations that fails to become detoxicated by incorporation in the uric acid molecule, or whether its occurrence in the excreta is of some positive significance in relation to excretory function, has not yet been determined.

d. Amino Acids

A variety of amino acids have been identified in the excreta of various insects, but the quantities involved are often so small that their occurrence could legitimately be regarded as reflecting a failure of resorption by the rectal epithelium (see Chapter 5) rather than as a normal part of the excretory process. In some insects, however, particularly among species which subsist on high protein diets, the quantities are substantial (see Table 6.1) and the distribution among different kinds of amino acids bears relation neither to the proportionate composition of the haemolymph, which would be reflected in that of the urine, nor to the proportionate composition of the diet, which might be reflected in that of digestive wastes, so that in these a selective elimination through the excretory system would appear to be involved. In the tsetse fly the main amino acids in the excreta are arginine and histidine, the two that have the highest nitrogen content of all protein amino acids, and their elimination could perhaps be seen as a method by which the metabolic losses involved in the synthesis of uric acid are minimized during the first half of the hunger cycle, when water reserves are plentiful.

e. Miscellaneous Materials

Pteridines occur in high concentration in the excreta of certain insects, but little is known of their excretory metabolism. Xanthine and hypoxanthine, reduced derivatives of uric acid (see Chapter 1), have been identified in the excreta of a few insects, sometimes making up as much as 10% of the total nitrogen. In blood-sucking insects haematin, or haematin derivatives, are conspicuous components of the excretory material.

2. The Storage of Excretory Products

The deposition of excretory products in various tissues of the body seems to be a regular feature of many species of insect. It is particularly common among members of the Orthopteran family Blattidae, in which as much as 10% of the total dry weight of the body may be uric acid. In many insects the uric acid is stored in the fat body, and in view of the readiness with which it may be mobilized from such deposits, and of the possibility that it may serve as a source of nitrogen for synthetic purposes, use of the term storage excretion may not be altogether appropriate. In other cases, however, the sequestration of uric acid seems to be permanent, as in its deposition in the cuticle of certain species; or the material may be destined for eventual elimination, as in certain cockroaches where large quantities of uric acid are deposited in the accessory sex glands, from which they are poured over the spermathecae during copulation; here a true disposal of nitrogenous wastes would seem to be involved.

CHAPTER 7

RESPIRATORY EXCHANGES

Perhaps the most striking feature of the organization of insects is the tracheal system. It represents in essence a system by whose means air is brought into the closest possible contact with sites of tissue respiration in all parts of the body. Air may be said to be piped to the respiring tissues, and their respiratory needs are thus taken care of in a very direct fashion. The insects are not unique in the development of such a mode of respiration, but it is in them that it reaches its finest expression, and the tracheal system of insects has attracted the attention of anatomists and physiologists alike, through the centuries.

The main function of the respiratory system is to enable an efficient exchange of the respiratory gases, oxygen and carbon dioxide, which feature respectively as an indispensable input and as an unavoidable output of the metabolic system. The fulfilment of this function is linked to the provision of a permeable membrane across which the respiratory exchange can take place, and a conflict of requirements arises in this connection. For it seems that, in most natural membranes, permeability to oxygen goes hand in hand with permeability to water, and a membrane across which oxygen can exchange freely is also one across which water can be readily lost to a dry atmosphere. This is a problem which faces all terrestrial animals, and in all it leads to a compromise in the sense that a balance is struck between the need to meet the demands of respiration on the one hand, and the necessity to limit respiratory exchange in the interests of water balance on the other. So closely interlinked are the processes of oxygen uptake and water loss, of respiration and transpiration, that they cannot sensibly be treated in isolation from one another. Unless the respiratory system is seen against the background of this compromise, many of its properties cannot be interpreted satisfactorily. What must be discussed in this chapter is, therefore, not only the exchanges of oxygen and of carbon dioxide, but also the losses of water which are unavoidably sustained in the process of respiration. Thus, while it is the respiratory system which is under consideration, discussion will not be limited to the exchange of respiratory gases.

Unfortunately the area of overlap between respiration and water balance is one that has been much neglected by experimental physiologists, who tend to fall into distinct categories depending on whether they are interested primarily in respiration, or primarily in water balance. As a result, a great deal of information may be available concerning details of respiratory function in one insect, and of the mechanisms by which transpiration is regulated in another, but there is no single instance of a concerted experimental attack on both aspects of the question in a single species. In order to present a coherent account, some play has unavoidably had to be given to speculation, in order to marry the results obtained from two quite distinct disciplines of physiological investigation.

1. The Structure of the Respiratory System

Before entering on a discussion of respiratory physiology, it will be necessary to give some account of the structure of the respiratory system, which is one of considerable complexity. Its main features are illustrated diagrammatically in Fig. 7.1, which indicates how oxygen is supplied to the flight musculature of an insect. The tracheal system opens to the thoracic surface at the spiracle, which is often set in a cuticular depression, and whose opening is usually guarded by a filter of cuticular bristles, serving to prevent entry into the tracheal system of

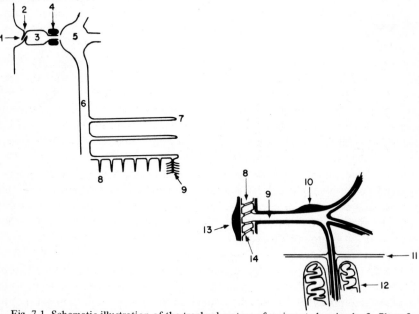

Fig. 7.1. Schematic illustration of the tracheal system of an insect. 1, spiracle; 2, filter; 3, atrium; 4, spiracular valve; 5, tracheal manifold; 6, primary trachea; 7, secondary trachea; 8, tertiary trachea; 9, tracheole; 10, tracheal end cell; 11, sarcolemma; 12, mitochondrion; 13, tracheal epithelium; 14, taenidium.

particulate material. An atrial cavity may be interposed between the filter and the spiracular valve; the valve itself takes a diversity of forms in different species of insect, but essentially it is a relatively simple system of cuticular levers operated by one or more bands of muscle, whose contractions cause the levers to bear on the tracheal tubes in such a way as to reduce their diameter to the point of total occlusion. Beyond the valve the trachea usually divides into a number of main branches, some of which establish communication with other parts of the tracheal system as longitudinal or transverse connectives. Further branching gives rise to secondary and tertiary tracheae of progressively smaller diameter and these in turn branch to form a series of fine tracheoles, which represent the final ramifications of the branched system, usually tapering from a diameter of $1.0\ \mu$ to end blindly at a diameter of 0.1-$0.2\ \mu$. Each tracheole lies within a single palmate cell, the so-called tracheal end-cell, whose nucleus is usually situated near the origin of the tracheole. The tracheolar branches are associated with a sheath, representing projections fom the surface of the end-cell.

In flight muscles the tracheoles often penetrate into the muscle cells themselves, carrying with them not only the sheath of the end-cell but also the invaginated membrane of the muscle cell. They branch within the muscle fibre to invest each intracellular myofibril, in the closest proximity to the longitudinal arrays of mitochondria. In other organs tracheoles do not usually penetrate the individual cells, and they may do no more than form a general surface investment of the tissue.

Developmentally the tracheal system represents an invagination of the epidermis, and each trachea is surrounded by a sheath of epithelial cells supported by a basement membrane, and lined with a substantial cuticular membrane (see Fig. 7.1). The cuticular intima is thrown into folds which run a spiral course to form the so-called taenidia, serving to strengthen the tubes against collapse. The tracheal lining is composed of layers similar to those of the surface cuticle, but in the tracheole the intima becomes extremely delicate, and its composition has not yet been determined.

The finest tracheolar branches are often partially filled with fluid, and the point to which the liquid column extends is thought to depend on a balance between the capillary force, tending to draw liquid out along the tube, and the colloid osmotic pressure exerted by the cytoplasm of the end-cell. It has been shown that an increase in haemolymph osmotic pressure, as during muscular exercise, particularly under conditions of low oxygen tension, is associated with absorption of fluid from the tracheoles, leading to an extension of air into the finer branches, as illustrated in Fig. 7.2. This would obviously serve to facilitate respiratory exchange between the tracheole and the site of respiration.

The distribution of the tracheal system varies enormously from species to species, and from stage to stage in the life history. In many insects each body segment bears a spiracle, and different segments are linked by lateral longitudinal trunks. There is, however, a general tendency towards reduction in the number of functional spiracles, and in some insects only the posterior pair of spiracles remains open, to serve the needs of the whole animal by way of the longitudinal trunks. Special thin-walled and collapsible dilatations, or air sacs, are a common feature of the tracheal system in many insects, occurring either as blind diverticula, or along the course of major tracheal routes. The precise distribution

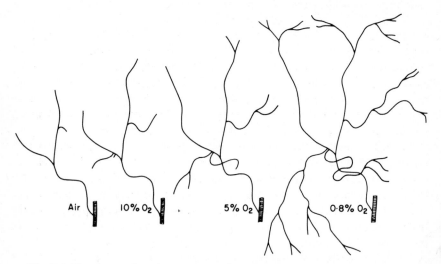

Fig. 7.2. The extent of gas in one group of tracheoles in the abdomen of a flea at rest and exposed to different concentrations of oxygen (Wigglesworth, 1965).

of the finer tracheae at any one stage in the life history may depend to a considerable extent on local demands during the previous stage, and if insects are maintained at low oxygen tensions during development, the tracheal investment of particular organs may be increased. Within a given developmental stage there is a tendency for tracheoles to migrate towards regions of low oxygen tension. In other words, while the gross pattern of tracheal distribution is determined genetically, there is scope for fine adjustment in relation to local needs, thus ensuring that the supply of oxygen to various tissues is related to the demand for oxygen by those tissues.

Attempts have been made to measure the total volume of the tracheal system of a number of different insects by a variety of methods, and values ranging from 5% to 50% of total body volume have been reported. The total number of tracheoles in a silkworm larva has been estimated as 1.5×10^6, and in view of the high surface to volume ratio which would characterize tubes of such fine dimension, there can be little doubt that they constitute the main site of respiratory exchange.

2. The Physiology of Respiration

There are two different ways in which respiratory gases can be transported in the tracheal system of insects—by bulk flow of air, or by diffusion in still air. Diffusion may be free in the sense that it is limited only by the fixed dimensions of the tracheal system, or it may be restricted by closure of the spiracular valves. For purposes of discussion it will be convenient to consider the physiology of respiration under the three corresponding headings of (a) diffusion in the open system, (b) diffusion in the regulated system and (c) ventilation.

a. Diffusion in the Open System

(i) Oxygen. The question whether the oxygen requirements of an insect can be met by diffusion of oxygen through the complex ramifications of the tracheal system is one which has occupied the attention of physiologists for several decades. Recent calculations have served to confirm the early estimates of Krogh, who in 1920 showed that diffusion is a major factor in the transport of oxygen through the tracheal system; and that only a small gradient of oxygen tension is required to account for the transfer of oxygen from the mouth of the tracheal tree to the site of tissue respiration, at a rate sufficient to meet the demands of oxygen consumption.

It is important to note that the rate at which oxygen diffuses in air is enormously greater than the rate at which it diffuses in water. The permeability constants are 11.0 and 0.00003 ml, $min^{-1}.,cm^{-2}.,atm^{-1}.,cm^{-1}$ respectively, implying that if a given gradient of oxygen tension suffices for the transport of oxygen along a 3-cm long tracheal tube, that same gradient would be adequate for a diffusion distance of no more than 0.1 μ in the aqueous medium of the cellular fluid. This emphasizes the need to bring the tracheolar supply into extremely close proximity to the site of tissue respiration, and Weis Fogh (1961) has calculated that during flight in the dragonfly, with the partial pressure of oxygen in the tracheoles at 142.5 mmHg, which is close to the 150 mmHg of the atmosphere, the maximum distance for tissue diffusion is about 10 μ. In other words, unless the mitochondria are within a distance of 10 μ from the nearest tracheole, the rate at which oxygen can be supplied to them would be insufficient to meet the demand. In actual fact, the tracheoles of active flight muscle are usually no more than 3 μ apart, so that the architecture of the system provides a substantial safety factor.

Such measurements as have been made of the tension of oxygen in tissue fluids and in the tracheal system of insects confirm that the drop in tension in the aerial phase is slight, probably of the order of 2% during rest, and 20-30% during flight, when the rate of transport has to be increased to meet the greater demand.

An attempt has been made to provide a diagrammatic summary of the situation in Fig. 7.3(a), where curve (i) represents the oxygen tension at

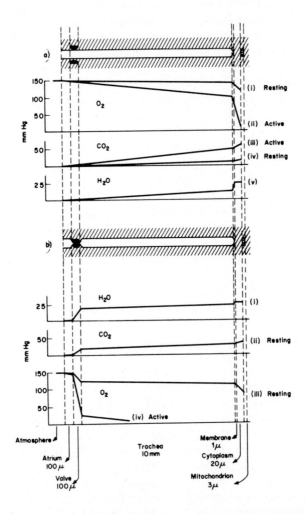

Fig. 7.3. Diagram to illustrate the disposition of tension gradients for oxygen, carbon dioxide and water in the unventilated tracheal system of an insect. (a) Spiracles open. (b) Spiracles closed. Since the cross-sectional area of the tracheal tree does not change substantially along its length, and since the bulk of respiratory exchange would take place through the relatively enormous surface area of the tracheolar branches, the system has been represented as a single unbranched tube extending between the atmosphere on the left and the site of tissue respiration, as represented by a mitochondrion, on the right. The linear scale has been greatly distorted, as indicated at the bottom of the figure, to allow graphical representation of the whole system. It is assumed that the respiratory gradients shown for the insect at rest are adequate to satisfy the needs of steady-state respiration at the rate of $x \, \mu lO_2/min$, and that respiratory rate during activity is 5 x; for further explanation see text.

different points along the diffusion path from the atmosphere on the left to the mitochondrion on the right, in an insect at rest. A very shallow gradient suffices for transport in the aerial phase, a much steeper one is required for transfer in the aqueous phase from the end of the tracheole into the cell and across to the mitochondrion, and it is at this point that most of the drop in tension occurs. Even so, the oxygen tension at the mitochondrial surface is represented as being relatively high, and reduced by no more than 20% from atmospheric.

During activity, or during exposure to high temperatures, when the rate at which oxygen is consumed would be substantially increased, the gradients would need to be correspondingly steeper to promote faster transport, as indicated in curve (ii) of Fig. 7.3(a). This can only be achieved by dropping the tension at the mitochondrial surface to a much lower level, one, however, that would still be adequate for maximal mitochondrial activity. In this case, again, the bulk of the gradient has to be exerted across the aqueous pathway.

(ii) Carbon Dioxide. The carbon dioxide which is produced at the site of tissue respiration will combine with water to give carbonic acid, and this, in turn, will dissociate to give the bicarbonate ion. The combination of carbon dioxide with water is a relatively slow reaction, and in the blood of vertebrates it is accelerated by an enzyme, carbonic anhydrase, present in the red blood corpuscles. This enzyme has not been demonstrated in the haemolymph of terrestrial insects, but it has been shown to be active in the tissue fluid of several species. This would be in accord with the fact that insect haemolymph plays little part in the transport of respiratory gases, and that the bulk of exchange occurs directly between tissue and tracheoles.

The bicarbonate content of insect haemolymph, and of the tissue fluids with which it is in equilibrium, is variable, and the relation between carbon dioxide tension and bicarbonate concentration (the CO_2-capacity) in insect haemolymph also differs widely between species. For present purposes a value of 5 m.eq/litre of bicarbonate in equilibrium with an atmosphere containing 2% carbon dioxide will be assumed, giving a tracheal carbon dioxide tension of about 15 mmHg. With the carbon dioxide content of the atmosphere at less than 0.1%, the gradient available for promoting the diffusion of carbon dioxide from tissue to environment would then be about 15 mmHg. This is very much less than that which promotes the transfer of oxygen in the reverse direction. The permeability coefficient for carbon dioxide in an aqueous medium, however, is about 36 times greater than that for oxygen, so that a much shallower gradient between mitochondrial surface and tracheole wall would suffice, while the gradient in the aerial phase would need to be very little greater, since the diffusion coefficient for carbon dioxide in air is only marginally smaller than that for oxygen. Conditions would thus be adequate for the disposal of carbon dioxide along a gradient system such as that illustrated in Fig. 7.3(a), curve (iv).

During activity carbon dioxide would be produced at a greater rate, and it would have to be disposed of correspondingly faster over a correspondingly steeper gradient system. The only way in which this could be achieved would be by raising the tension of carbon dioxide at the tissue level, and thus it has been represented in curve (iii) of Fig. 7.3(a); it should be mentioned that experimental evidence for an increase in carbon dioxide tension during activity is lacking.

Since carbon dioxide can diffuse more rapidly than oxygen through water and through lipids, the proportion of carbon dioxide which exchanges through the general body surface, rather than through the tracheal system, would be proportionately greater. In experiments with insects whose spiracles have been blocked, substantial diffusion of carbon dioxide takes place through the general body surface, but it is likely that this would be associated with abnormally high internal tensions of carbon dioxide. Under normal conditions the fraction that exchanges through the body surface is probably small, and it may be ignored in the present context.

(iii) Water Vapour. No careful theoretical approach has yet been made to the problem of the disposition of gradients of vapour pressure within the tracheal system of insects, nor are experimental data available that could serve as a basis for a theoretical model. The account which follows will therefore be superficial and speculative, and will have to be regarded as no more than a first rough approximation.

The total gradient available to promote the diffusion of water vapour from the insect to the environment is set, on the one hand, by the saturated vapour pressure of air in equilibrium with the tissue fluids, and on the other, by the water vapour pressure of the atmosphere. With an insect exposed to dry air at a temperature of 25° it would amount to about 25 mmHg, with the gradient extended between the tissue fluids in contact with the tracheolar wall and the mouth of the tracheal tree. The tracheoles appear to be lined by a layer of cuticulin (see Chapter 1) constituted by a tanned lipoprotein complex. No information is available concerning the permeability of this layer, but it seems reasonable to suppose that it would offer substantial resistance to the diffusion of water, and so it has been represented in curve (v) of Fig. 7.3(a); approximately half of the total gradient is represented as acting across the tracheolar membrane. In this case the disposition of gradients would, of course, be the same in resting as in active insects, except in so far as the rise in temperature associated with activity would slightly increase the saturated vapour pressure.

b. Diffusion in the Regulated System

(i) Water Vapour. The rate at which water is lost from the tracheal system under the conditions illustrated in Fig. 7.3(a), that is, with spiracles open, has been determined for a number of insects, and has been demonstrated to be

substantial. This in itself indicates that a considerable gradient of vapour pressure is acting across a relatively permeable respiratory membrane, so that the satisfaction of respiratory requirements may be seen as posing a threat to water balance. The way in which this threat has been countered in most classes of insects is by the development of a mechanism for closing the spiracle, in other words, by the interposition of resistance to diffusion at one point in the diffusion pathway, as illustrated in Fig. 7.3(b). Since diffusion would be greatly impeded by occlusion of the tracheal lumen, it is clear that a great part of the available gradient would under these circumstances have to be exerted across the region of resistance, as illustrated in curve (i) of Fig. 7.3(b). As far as water vapour is concerned, the result is that a relatively small proportion of the total gradient is now exerted across the respiratory membrane, and the rate at which water is lost through that membrane, and hence from the system as a whole, will be correspondingly reduced. Quantitative aspects of this effect are illustrated in Fig. 7.4, which shows the water loss of tsetse flies at different relative humidities. In air containing 15% carbon dioxide the spiracles are kept permanently open, the rate of water loss is high and directly proportional to relative humidity (Fig. 7.4, curve i), as it would be in a simple physical system. The rectilinear relation is maintained when the spiracles are artificially blocked, but under these circumstances the rate of water loss is greatly reduced (Fig. 7.4, curve iii), representing transpiration through the general body surface. With the

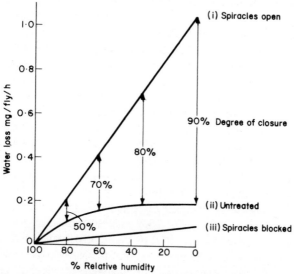

Fig. 7.4. The rate of water loss of tsetse flies at different relative humidities. (i) In air containing 15% carbon dioxide. (ii) In air. (iii) In air with spiracles blocked. The degree of effective closure of the spiracles is indicated at four different levels of humidity; for further explanation see text (schematized from Bursell, 1957).

normal insect in air free of carbon dioxide, the rate of water loss is intermediate between these two extremes, and its relation to relative humidity shows a marked departure from rectilinearity, with water loss strongly reduced at low humidities (Fig. 7.4, curve ii). The difference between curves (i) and (ii) at any one point, expressed as a proportion of the difference between curves (i) and (iii) at that same point, gives a measure of the degree to which the spiracles are closed, and values have been included at representative points of the curve. They show that in dry air the effective mean spiracular aperture is reduced by 90%, while in humid air the extent of occlusion is very much less. Whether these values indicate that the spiracles are held at the corresponding degree of closure, or whether they indicate that the spiracles are kept closed for the corresponding proportion of time, has not yet been determined; the subject will be discussed further in Chapter 10, where neurophysiological aspects of spiracular regulation will be considered.

(ii) Carbon Dioxide. The distribution of gradients for carbon dioxide will be affected in a way similar to those for water as a result of spiracular occlusion, and to ensure the disposal of carbon dioxide at the rate at which it is produced by the resting insect, the carbon dioxide tension at the site of respiration would need to be raised to more than twice its former value, as indicated by curve (ii) of Fig. 7.3(b).

(iii) Oxygen. With oxygen, too, a considerable proportion of the available gradient will have to be extended across the spiracular valve, so that, to maintain a transfer at a rate equal to the rate of consumption, the tissue tension would have to be dropped from about 120 mmHg to about 90 mmHg (see curve (iii) of Fig. 7.3(b)). The diagram thus illustrates in more specific terms the existence of a conflict between the requirements of water balance and of respiration, in so far as it shows how conditions which favour the supply of oxygen (open spiracles) favour also the loss of water, while conditions that reduce losses of water (closed spiracles) will jeopardize respiratory requirements to some degree.

It is clear from Fig. 7.3(b) that under the hypothetical conditions illustrated, the tension of oxygen at the site of respiration is still well above that which is required for maximal mitochondrial activity. It is not until the active animal is considered that the conflict finds full expression. On the assumption that oxygen consumption increases five-fold, the gradient across the spiracle would need to be five times as steep in order to supply oxygen at the requisite rate, and so would the gradient along the main tracheal pathways. Gradients of this magnitude are illustrated in curve (iv) of Fig. 7.3(b), which may be said to run out of oxygen tension long before the end of the pathway is reached, indicating that steady-state respiration at this rate would be impossible. The closure of spiracles would obviously militate against even a moderate rise in metabolic rate, to say nothing of the 50- to 100-fold increases which may characterize insects in flight. It is in fact a matter of common observation that bursts of struggling in restrained insects are usually associated with momentary opening of the spiracular valves, while the onset of flight activity is accompanied by sustained opening. This suggests that during flight the requirements of respiratory exchange may take precedence over requirements of water balance; the spiracles are opened wide to permit access of oxygen, while water is allowed to be freely lost from the respiratory membranes. It is possible, however, that the conflict may be more finely balanced than that, even during flight. The spiracles might be opened no more than would satisfy the requirement for oxygen, so that although the rate of water loss is still high, it may not be completely unregulated. Very little work has been done on the water loss of flying insects, and such a possibility is by no means excluded by available evidence.

c. "Discontinuous Respiration"

The phenomenon of spiracular regulation was brought forcefully to the attention of respiratory physiologists during the 1950s, when anomalous results were obtained by a number of workers, who were investigating the respiratory exchange of quiescent or inactive insects. It was observed that, while the oxygen consumption proceeded continuously at a steady rate, the release of carbon dioxide occurred in intermittent bursts, as illustrated in Fig. 7.5(b). This rather puzzling phenomenon became tbe object of a series of intensive investigations by American workers, who used the diapausing pupae of certain large moths as convenient experimental material, and a satisfactory interpretation of the phenomenon of "discontinuous respiration" has been proposed on the basis of this work.

It was shown, in the first place, that the cyclical release of carbon dioxide could be abolished by intubation of one or more of the spiracles, indicating that the spiracular valves were implicated. This was confirmed by observations of the exposed spiracular valves during the burst cycle. Spiracles were kept firmly closed immediately following a carbon dioxide burst; a period then followed during which slight fluttering of the valves could be observed (see Fig. 7.5(a)), until eventually the spiracles would open wide to complete the cycle with the release of another burst of carbon dioxide.

Measurements of the intratracheal pressure and of the tension of oxygen and carbon dioxide showed that, after a burst, the oxygen tension drops steeply from atmospheric levels down to about 25 mmHg, at which level it is maintained until the next burst (see Fig. 7.5(c)). The intratracheal pressure drops slightly during the interburst period, while the tension of carbon dioxide rises slowly from about 25 mmHg at the end of a burst to 50 mmHg at the end of the interburst, falling sharply as soon as the spiracles open.

Results of this kind have served as the basis for an hypothesis of "discontinuous respiration" put forward by Buck (1958), who envisages the following sequence of events. At the end of a carbon dioxide burst the tracheal

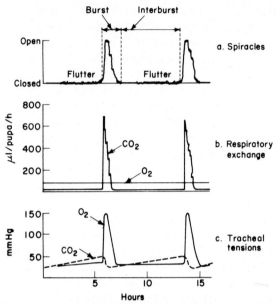

Fig. 7.5. Respiratory exchange in the diapausing pupae of *Hyalophora*. (a) The degree of closure of spiracular valves at different points in the cycle of discontinuous respiration. (b) The release of carbon dioxide and the consumption of oxygen during discontinuous respiration. (c) The tensions of carbon dioxide and of oxygen in the tracheal system during discontinuous respiration. Based on Schneiderman and Williams (1953), Schneiderman (1960) and Levy and Schneiderman (1958).

system has been flushed with air, so that oxygen concentration is high and carbon dioxide concentration relatively low. Consumption of oxygen by the respiring tissues leads to a fall in oxygen tension, the supply of oxygen by diffusion from the atmosphere being at this stage prevented by the closed spiracular valve. The carbon dioxide which is produced during respiration is to a large extent held in solution as bicarbonate; since it does not appear as a gas, to balance the disappearance of oxygen, the net result is a fall in pressure within the tracheal system. As the amount of bicarbonate, and to a smaller extent of the gaseous carbon dioxide with which it is in equilibrium, builds up, the spiracles open minutely and intermittently during the period of "fluttering". Since the intratracheal pressure is less than atmospheric, there will be a bulk flow of air, containing 21% oxygen, into the tracheal system. As the oxygen continues to be used, the low intratracheal pressure will be maintained, ensuring a sustained in-flow of air, and a continuous replenishment of oxygen. With the progressive utilization of oxygen, the intratracheal gas becomes progressively enriched with nitrogen, and carbon dioxide tensions will also increase steadily until a critical point is reached, when a new burst is initiated and the spiracles open wide. At this stage there would be rapid equilibration between the atmosphere and the tracheal system, oxygen diffusing in and nitrogen and carbon dioxide out along their respective gradients of tension. The carbon dioxide held as bicarbonate would be released as a result of the drop in intratracheal carbon dioxide tension. At the end of the period of equilibration, the spiracles close and the whole cycle is repeated.

It is a measure of the degree of specialization within the field of insect physiology, to which reference has already been made, that the phenomenon of "discontinuous respiration" has been considered and interpreted in all but complete isolation from its *raison d'être,* which can hardly be other than water balance. It is difficult to conceive what respiratory benefit could arise from the use of a system that involves a progressive fall in tracheal oxygen tension and a progressive rise* in carbon dioxide tension, and there appears to be general agreement that, as with other aspects of spiracular regulation, the system must find its ultimate interpretation in the context of water balance, but as yet there has been no experimental attack on this aspect of the problem. The theoretical treatment of Buck (1958) has indicated that the outward diffusion of water vapour would be impeded by the bulk inflow of air during the interburst period, so that the system is likely to be effective in the conservation of water. But there has as yet been no attempt to determine experimentally what the quantitative implications of discontinuous respiration are in terms of water balance; indeed, the only experimental evidence that it is related to water balance at all is circumstantial, namely that pupae maintained at high relative humidities release carbon dioxide continuously.

d. Ventilation of the Tracheal System

The diffusion of respiratory gases between the atmosphere and the site of respiration is adequate to account for the respiratory exchange of many insects both during rest and in flight. In some, however, and particularly among the larger species, the diffusion path appears to be too long for respiratory requirements to be wholly satisfied in this way even in the resting insect, and in most, the process of diffusion becomes inadequate during the periods of intense metabolic activity which are associated with flight. In these a process of tracheal ventilation is superimposed on diffusion processes, oxygen within the tracheal system being replenished, and carbon dioxide flushed, by bulk flow of air in the larger tracheal branches. Diffusion is still the process by which respiratory gases are exchanged between the primary tracheae and the site of tissue respiration, and the main function of ventilation is simply to shorten the path of diffusion.

(i) Ventilation in the Resting Insect. Tracheal ventilation in the resting locust has been the object of a series of detailed investigations by Miller, the results of which form the basis of the following account. Four types of ventilatory movement have been distinguished, one involving the raising and lowering of the floor of the abdomen, one involving a telescoping of abdominal segments, and

two based on protraction and retraction of the head and of the prothorax respectively. The last three should be considered essentially as auxiliary mechanisms, which come into play only for short periods when the demands of respiratory exchange are intense.

The pressure exerted by muscular movements of the type described on the sprung walls of the larger tracheae cause alternate compression and expansion, leading to corresponding displacements of the air that they contain. In the resting insect about 86% of the total ventilation is associated with respiratory movements of the abdomen, with thoracic and cephalic components contributing no more than 11% and 3% respectively. Only about 5% of the total volume of the tracheal system is exchanged with each pumping stroke under normal conditions, but during hyperventilation the value may rise to 20%, and in other species as much as 70% of the total volume can be ventilated.

The frequency and ampltude of abdominal ventilation can be recorded very simply on a smoked drum by attaching a light lever to one of the abdominal sternites of a locust, restrained ventral surface uppermost by strips of plasticene. Examples of such records are shown in Fig. 7.6, where traces (a) and (b) show that the intensity of ventilation is greatly increased when the cephalic or the thoracic nerve centres are exposed to increased concentrations of carbon dioxide. Abdominal nerve centres are capable of initiating respiratory movements in their own segments, but these autonomous movements are of lower frequency and amplitude, and are insensitive to carbon dioxide. It appears that under normal conditions their rhythm is subject to the overriding influence of carbon dioxide sensitive pacemakers situated in higher centres.

The intensity of ventilation is also affected by oxygen lack, but at a much lower level of sensitivity; it is necessary to drop oxygen tensions to about half of their normal value to produce the same degree of hyperventilation as is produced by 1% carbon dioxide.

Earlier work had shown that the activity of spiracular valves was closely synchronized with ventilatory movements, in such a way as to produce a directed stream of air through the main tracheal trunks. This was confirmed by a detailed study of the behaviour of each of the 10 spiracles, and the results have been summarized in Fig. 7.6 (c) and (d). In the resting locust most of the

Fig. 7.6. Aspects of tracheal ventilation in the locust. (a) Tracing of kymograph records of the abdominal ventilation in the locust; the arrow marks the time at which air containing 6% carbon dioxide was injected into the cephalic tracheal system. (b) As for (a); the arrow marks the time at which the metathoracic ganglion was perfused with saline equilibrated with 5% carbon dioxide. (c) Schematic representation of the air flow resulting from ventilation during rest. Spiracles are numbered from 1 to 10, and the direction of flow is indicated by arrows. a.s., pterothoracic air sac. (d) Schematized representation of the closure of spiracular valves in relation to the phases of abdominal ventilation. I, inspiration; E, expiration. (e) Tracings of kymograph records of the abdominal ventilation in resting locust starved for 27 hr in dry air. (f) as for (e); after 27 hr starvation at high relative humidity. (a)-(d) from Miller, 1960; (e)-(f) from Loveridge, 1968.

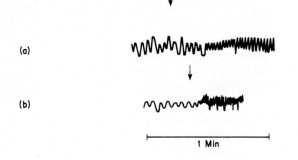

(a)

(b)

1 Min

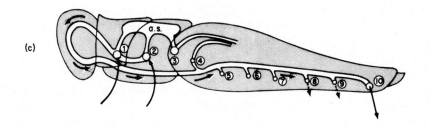

(c)

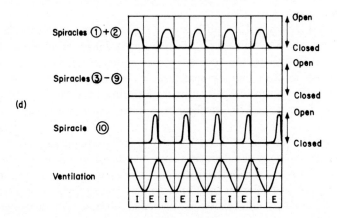

(d)

Spiracles ①+②

Spiracles ③-⑨

Spiracle ⑩

Ventilation

Open

Closed

Open

Closed

Open

Closed

I E I E I E I E I E I E

(e)

(f)

spiracles remain closed during all phases of the cycle of ventilation. Spiracles 1 and 2 open during early parts of the inspiratory phase, at which time all other spiracles are closed, so that air is drawn in through the first two spiracles. During expiration spiracles 1 and 2 close, and spiracle 10 opens, allowing air to escape through the posterior spiracle, and giving a flow through the main tracheal trunks as illustrated in Fig. 7.6(c). Under conditions of greater oxygen demand, other abdominal spiracles may be brought into use for expiration, in sequence from the posterior end.

It has been demonstrated recently that the intensity of ventilation in the locust is regulated in relation to the demands of water balance. During exposure to desiccating conditions there is a progressive reduction in the amplitude and frequency of abdominal ventilation (see Fig. 7.6 (e) and (f)). This is undoubtedly another reflection of the conflict that exists between respiratory requirements and the need to limit transpiratory losses of water. By reducing the rate of flow of air through the primary branches of the tracheal system the vapour pressure gradient between the mouth of secondary tracheae and the tracheolar membranes is reduced, thus decreasing the rate of water loss across the respiratory surface. By the same token, the gradient of oxygen tension would be decreased, so that the oxygen tension at the site of tissue respiration would need to be dropped in order to ensure a supply of oxygen commensurate with the demands of respiration; and similarly, the carbon dioxide tension would need to be increased in order to maintain the requisite gradient for disposal at the requisite rate.

Details of the ventilatory pattern differ substantially from species to species; in some the flow may be predominantly tidal, and directional air streams may be weakly developed; in others the air may enter posterior spiracles during inspiration and leave through thoracic spiracles. Whatever the precise pattern, ventilation is invariably confined to the larger tracheal tubes, and the transfer of gases in the finer branches of the tracheal tree must always take place by diffusion.

(ii) Ventilation During Flight. To meet the tremendous increase in the demand for oxygen during flight there is a marked change in the pattern of ventilation. In the locust this is reflected particularly in the behaviour of spiracle 2 and the bringing into operation of spiracle 3. At the onset of flight both these spiracles open fully, and in early phases of flight they remain open during all phases of abdominal ventilation, the frequency and amplitude of which shows a pronounced increase. The opening and closing of spiracle 1 remains in phase with abdominal ventilation, the only difference being that during early phases of flight it tends to open more fully than it does during rest. In addition, all of the abdominal spiracles take on the pattern of activity of spiracle 10 during rest, serving, that is, as additional points of exit for the air stream entering the anterior spiracles. During later phases of flight there is a tendency for spiracles 2

and 3 to close during abdominal inspiration, so that the contribution of spiracle 1 to the directional flow of air through the cephalic system and backwards to the abdomen will tend to increase. The over-all effect of the changed pattern of ventilation will thus be to increase the flow of air which characterizes the resting insect, and to superimpose on this a tidal flow to the flight musculature of the pterothorax, in and out of the thoracic spiracles.

Pterothoracic ventilation in the flying locust has been the object of detailed study by Weis Fogh (1961), who showed that while abdominal ventilation would be quite inadequate to provide for recorded consumption during flight, locusts with the abdomen completely removed were able to fly without sign of respiratory distress. It seems that not only is the pterothoracic tracheal system in some degree anatomically isolated from the rest of the tracheal system (see Fig. 7.6(c)), but it is capable of acting as an autonomous unit of ventilation. The tidal flow is brought about in part by small amplitude movements of the pterothoracic walls and not associated with the wingstroke, and in part by contraction of the wing muscles themselves, which ventilate the proximal parts of the secondary tracheae. The volume change during the wingstroke is quite small, but at 1040 strokes per min it delivers 750 litres of air per kg of flight muscle per hr, which is well above the level of requirement for the metabolic machinery, at about 600 litres/kg/hr.

3. Conclusion

It has been indicated that insects have solved the problem of respiratory exchange in a direct and very simple way, by bringing air rich in oxygen and poor in carbon dioxide into the closest juxtaposition to the site of respiration. But while the basic principle of tracheal respiration is simple, the mechanism by which it is achieved has in most insects become greatly elaborated in order to meet the demands of water conservation. This has been achieved by the development of spiracular valves which ensure, in one way or another, that the water loss associated with respiratory exchange is reduced to a minimum. The development of mechanisms for closing the spiracles has in turn opened the way to the production of a directional flow of air through the main tubes of the tracheal system, so that respiratory demands can, where necessary, be met in part by bulk flow rather than solely by diffusion.

SECTION II

Neuromuscular Physiology

INTRODUCTION

In Section I the satisfaction of metabolic requirements has been considered in terms of somatic physiology, the processes of which ensure the requisite input of food materials and oxygen, the elimination of waste products and the maintenance of conditions appropriate to the proper functioning of the metabolic machinery. In that section the insect was considered largely as a self-contained entity, and the satisfaction of different metabolic requirements was discussed with reference to the corresponding organ systems. But an insect should obviously not be regarded simply as the sum total of a number of autonomous systems, since the continued existence of the whole depends very much on an appropriate co-ordination between the activities of its different parts. Nor is it legitimate to consider any metabolic system as self-contained and isolated from the environment which sustains it. The satisfaction of the requirement for food, for instance, involves a particularly complex interaction between the insect and its environment. At this stage attention must therefore be turned to the mechanisms by which co-ordination between the activities of different organ systems is achieved, and to the means by which the insect is enabled to respond appropriately to features of the environment which are of importance in relation to its continued existence; and it is at this point, therefore, that the properties and functions of the neuromuscular system of insects must be examined. The same approach will be followed as before, in that attention will be focussed primarily on those features of the neuromuscular physiology of insects which appear to set them apart from the generality of animals. The conduction and transmission of the nerve impulse, the contraction of muscle and the initiation of afferent input at the level of the sense organ will be considered first, as what may be termed the unit processes of nervous function. After that the integrative aspects of nervous function will be discussed, firstly in terms of the activity of component units and their interaction in isolated systems, and secondly as it is manifested in the complex behaviour of the insect as a whole, seen against the background of its neurophysiological basis. Such an approach may seem ill-advised at the present time, since our knowledge of neurophysiology is by no means adequate to provide a firm foundation for the interpretation of behaviour, nor is it likely that insects will prove particularly favourable material for an eventual bridging of the gap between the two disciplines of investigation. It has, nevertheless, been adopted in preference to a purely empirical approach on the grounds that, though it cannot hope to provide a convincing interpretation of behaviour, yet it may give some indication of the general terms in which such an interpretation must ultimately be sought.

CHAPTER 8

NERVES AND MUSCLES

1. Conduction of the Nerve Impulse

The nature of the nerve impulse and the mechanism of its conduction along the nerve fibre appear to be the same in insects as in all other animals that have been investigated, and capable of interpretation in terms of the ionic hypothesis, as developed on the basis of studies with other animal groups. According to this hypothesis the capacity for conduction should be markedly influenced by the proportionate concentration of inorganic ions in the medium that bathes the nerve fibre; high concentrations of potassium in the extracellular fluid should tend to reduce the magnitude of the resting potential, while low concentrations of sodium should reduce the positive overshoot of the action potential, on which the propagation of the impulse depends. These general properties of excitable tissue are of particular interest in relation to nervous function in insects, in view of the wide variation in haemolymph composition which characterizes different members of the group (see Chapter 5). In most species the Na^+/K^+ ratio is substantially different from the value of about 10 which characterizes vertebrate body fluids, and in some of the phytophagous members the concentration of potassium and magnesium may actually exceed that of sodium, yet nervous function seems to be in no way impaired.

The occurrence of a characteristic sheath investing the nerves and ganglia of insects was seen to be of possible significance in relation to this problem, and the structure of the sheath and of its associated cellular elements has come under intensive investigation. The sheath itself, called the neural lamella, constitutes a substantial investment measuring as much as 5μ in thickness, and it is made up of a number of distinct layers, as shown in Fig. 8.1(a). These include an outer amorphous region, a narrow fibrillar layer in the middle, and an inner layer in which tangentially disposed fibres of a collagen-like material lie imbedded in a homogeneous matrix. Underneath the neural lamella is a cellular layer called the perineurium, the cells of which are characterized by massive accumulations of glycogen granules, and by the occurrence of clusters of elongated mitochondria.

In view of the fibrous structure of the neural lamella there can be little doubt

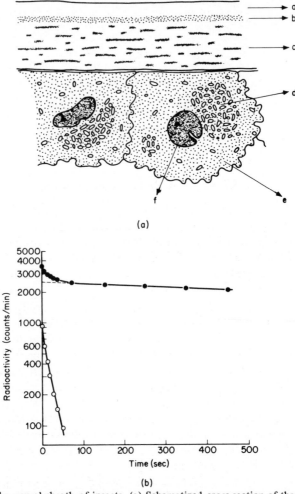

(a)

(b)

Fig. 8.1. The neural sheath of insects. (a) Schematized cross-section of the nerve sheath and perineural layer in the cockroach. a, amorphous layer; b, zone of fine filaments; c, fibrils of collagen-like substance in the main region; d, clusters of mitochondria; e, granules believed to be glycogen; f, nucleus of perineural cell. (Drawn from electron micrographs of Smith and Treherne, 1963.) (b) The escape of ^{22}Na from the nerve cord of the stick insect during washing with non-radioactive physiological saline. Closed circles, total exchange; open circles, fast component of the total exchange, obtained by subtraction of exponential portion extrapolated to zero time (Treherne, 1965a).

that it serves the function of mechanical support for the nervous system which it invests. The occurrence, however, of a cellular layer, with a submicroscopic structure that suggests a high level of metabolic activity, would indicate that the complex plays a more active role than that of support. The possibility that it might function to maintain the inorganic composition of extracellular fluids at a level suitable for nervous function has been investigated by Treherne, whose work forms the basis of the discussion that follows. He demonstrated first of all that inorganic and organic ions exchange quite readily between central nervous system and haemolymph. The exchange occurs as a two-stage process, with an initial rapid phase giving way to a slow, exponentially decaying, exchange. This is shown in Fig. 8.1(b), where the loss of radioactive sodium, from a nerve-cord previously loaded with the isotope, is depicted as it occurs during washing in non-radioactive saline. The evidence suggests that the initial flux represents an exchange between haemolymph and the extracellular fraction of the nervous system, and determination of the corresponding extracellular volume has enabled estimates to be made of the composition of extracellular and intracellular fluids for comparison with that of the bathing fluid. It can be seen from Table 8.1 that there are marked differences between the external medium and the extracellular compartment, and in both species the extracellular fluid shows an elevated Na^+/K^+ ratio. In the cockroach, the difference can be accounted for on the basis of a simple Donnan equilibrium between external and extracellular compartments across the nerve sheath; this is the reason why removal of the nerve sheath results in a change in the composition of extracellular fluids, and why de-sheathed preparations are rapidly depolarized by high potassium concentrations in the external medium, while in normal preparations the rate of depolarization is much lower. In the stick insect, however, the distribution of cations between the two compartments cannot be interpreted on the basis of a simple Donnan equilibrium. Here the relatively high concentration of sodium in the extracellular compartment was found to suffer substantial reduction in the presence of metabolic inhibitors, and it was concluded that an active process was involved in the maintenance of this difference in concentration across the neural sheath.

Despite the existence of such an active process, the ionic ratio of the extracellular fluid is far from being equivalent to that of the body fluids in other animal groups, and the question remains whether its composition is in fact compatible with the conduction of nerve impulses as interpreted by the ionic hypothesis. On the basis of the equilibrium potentials for potassium and sodium, as set out in the last column of Table 8.1, it can be seen that the conditions for impulse conduction along the lines of classical membrane theory would, in fact, be fulfilled, albeit on the basis of a rather low level of resting potential (-37 mV as compared with -64 mV for the cockroach and -70 mV for vertebrates in general), and with the potentiality for positive overshoot at the peak of the action potential, as given by the sodium potential, substantially reduced ($+22.3$ mV as compared with $+35.8$ mV for the cockroach and $+60.0$ mV for vertebrates).

It may be concluded that the apparent anomaly posed by the specialized

TABLE 8.1

The distribution of sodium and potassium between the external medium and extracellular and intracellular compartments of the nerve cord of the cockroach and the stick insect (from Treherne, 1965b)

		mmoles/l		mV Equilibrium potential
	External	Extracellular	Intracellular	
Cockroach				
Na^+	158	284	67	+35.8
K^+	12	17	225	−64.0
Na^+/K^+	13	17	−	
Stick Insect				
Na^+	20	212	86	+22.3
K^+	34	124	556	−37.1
Na^+/K^+	0.6	1.7	−	

composition of the haemolymph in certain groups of insects may in all probability be resolved on the basis of Treherne's investigations. The nerve axons themselves appear to be maintained in an environment which is compatible with impulse conduction by virtue of the presence, and in some cases of the activity, of ensheathing elements.

2. Neuromuscular Transmission

The small size of insects would be likely to raise a special problem in the context of neuromuscular activity, because many of the muscles on whose activity movement is based are minute, and the reduction in size of the motor apparatus is achieved by a reduction in the number of muscle fibres rather than by a reduction in the size of the fibres. The problem of producing a finely graded contraction, on which well co-ordinated movement must depend, is therefore not capable of solution on the basis of the familiar vertebrate pattern, where each muscle is divided into a number of motor units, and the contraction of the muscle as a whole is based on the recruitment of motor units, each of which contracts maximally. In insects, where muscles may contain no more than one or two fibres, a mechanism must be provided whereby the contraction of each individual fibre can be graded over a wide range, and interest will centre on the means by which this is achieved.

a. Anatomy of the Motor End-Plates

Certain anatomical features of muscle innervation in insects are of obvious relevance in relation to the problem of the graded response. In the first place, each muscle fibre is provided, not with a single end-plate as in vertebrate fibres, but with a series of end-plates, distributed at regular intervals over the whole surface of the fibre, with a spacing of between 40 and 80 μ, and with each motor end-plate supplied by its own branch of the motor nerve. Secondly, a proportion of the muscle fibres in the muscles of many insects are dually innervated, each neuromuscular junction supplied by two separate axons, as illustrated in Fig. 8.2. One of the axons is designated as the "slow" axon, the other as the "fast", on the basis of their physiological performance which will be described below. Whether singly or dually innervated, the motor end-plate typically takes the form of a claw of branching axon-terminals, sunk into gutters on the surface of

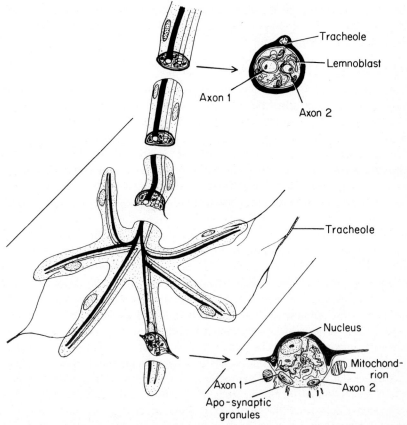

Fig. 8.2. Schematic drawing of the structure of a typical insect neuromuscular junction. The two motor axons travel together inside a common sheath, from which they both emerge at the point of contact with the muscle fibre, where the basement membrane of the muscle fibre fuses with the neural lamella of the nerve. The terminal branches fit into simple grooves on the surface of the muscle fibre, accompanied by the cells of the neural sheath (Hoyle, 1965).

the muscle cell, often accompanied by tracheolar branches. The axon membrane comes into intimate contact with the muscle membrane, and synaptic vesicles have been shown to be associated with the presynaptic membrane. By analogy with the vertebrate pattern it may be presumed that these presynaptic vesicles contain the neuromuscular transmitter.

A third axon has been shown to innervate a proportion of fibres in the tibial extensor of the jumping leg of the locust, and the possibility that this axon may be comparable to the inhibitory fibre of the Crustacean system has been considered. It has not so far been possible to correlate activity in this axon with unequivocal mechanical effects in the isolated preparation, but if activity is monitored in unrestrained animals it is found to coincide with flexion of the tibia, lending support to the view of an inhibitory function. However, detailed discussion of the physiology of this fibre would be out of place, in view of the uncertainty concerning its precise role.

b. Physiology of the Motor End-Plates

The physiology of neuromuscular transmission appears to be basically the same in insects as in other animals, involving a release of transmitter substance from presynaptic nerve terminals, triggered by the arrival of impulses in the motor nerve. Miniature end-plate potentials have been detected in unstimulated insect preparations, and by analogy with the situation in vertebrates, they have been interpreted as caused by the spontaneous release of transmitter quanta from the presynaptic membrane. The difference between insects and vertebrates appears to lie not so much in the mechanism of transmission, as in the nature of the transmitter. While acetyl choline has been firmly implicated as the transmitter of the vertebrate neuromuscular junction, attempts to establish the occurrence of cholinergic transmission in insects have proved unsuccessful. It is only recently that a likely candidate for the role of transmitter has appeared, in the form of glutamic acid. This substance is capable of causing depolarization of the motor end-plate in insects at physiological concentrations, and it appears that presynaptic stimulation causes the liberation of glutamic acid from *in vitro* preparations. It is a little surprising to find an amino acid of such central metabolic importance (see Chapter 1), and a substance so widely distributed and occurring in such high concentrations in haemolymph and muscle fluids alike, involved in the special role of neuromuscular transmitter. The apparent anomaly cannot be accounted for on the basis of a sealing-off of receptive regions from the haemolymph, since the presence of glutamic acid in low concentration is capable of causing tonic contraction in perfused preparations, indicating that the substance gains ready access to the receptor surface. The possibility that the concentration of "free" glutamate in the haemolymph may be much lower than would appear from standard amino acid analyses has been raised, but experimental evidence is lacking.

(i) The Slow Response. Stimulation of the slow axon with single shocks causes depolarizations of the muscle membrane ranging in magnitude from 2-30 mV, depending on the particular site from which recording is made. Little mechanical activity is associated with single shock stimulation, but if the preparation is stimulated with a train of impulses at the rate of 5-20 stimuli/s, the end-plate potentials show considerable facilitation (Fig. 8.3(a)), and a slow

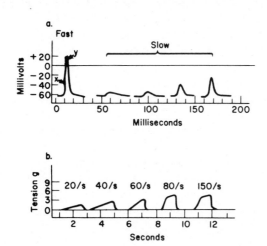

Fig. 8.3. Slow and fast responses of the tibial extensor muscle of the locust. (a) Electrical responses: stimulation of the fast axon gives a large, non-facilitating end-plate potential (x) on which is superimposed an active membrane response (y); stimulation of the slow axon gives a small end-plate potential, which shows a progressive increase in size during repetitive stimulation, but does not initiate an active response. (b) Mechanical response of the slow muscle fibres: with increasing frequency of stimulation, as indicated above each trace, the velocity and extent of contraction increases. (Redrawn from Hoyle, 1965.)

mechanical response is evoked. The greater the frequency of stimulation the greater, usually, the degree of facilitation; the development of tension is correspondingly faster and the peak tension attained greater, up to a limit at about 80 stimuli/s (Fig. 8.3(b)).

(ii) The Fast Response. Stimulation of the fast axon with single shocks produces a large end-plate potential with an electrically excited spike component (Fig. 8.3(a)), and the extent of depolarization is not facilitated by repetitive stimulation. The positive overshoot is not of sufficient magnitude, in the normal fibre, to ensure all-or-none propagation, so the response is conducted decrementally from the motor end-plate. The mechanical response to a single shock is a well-developed twitch; the twitch contractions fuse at about 10 stimuli/s, but the mechanical response is greatly facilitated at higher frequencies, to give tetanus/twitch ratios of as much as 10.

c. The Graded Response

On the basis of the anatomical and physiological results discussed, one can visualize, in general terms, how insects have solved the problem of producing an accurately graded response from muscles which comprise a minimum of fibre units. The basic approach has been through a graded electrical and mechanical response, as opposed to the all-or-nothing response that characterizes the vertebrate system. If we consider a muscle like the extensor of the locust jumping leg, it is clear that a full range of tensions from zero to maximal can be elicited by suitable patterns of activation in the slow and fast axons, with the lower range of tensions and contraction velocities catered for by tetanic discharge of the slow motor neurone at increasing frequencies, the higher range by superposition of fast axon activity, operating on the basis of mechanical facilitation at peak tension. Since all electrical events are essentially local, the multi-terminal innervation is a necessary condition for maximal activation of the contractile machinery. It should be mentioned that in a number of insect muscles, some opportunity for gradation based on subdivision of the muscle into motor units exists, side by side with the capacity for gradation in individual units. The tibial flexor of the locust, for example, is innervated by three fast motoneurones which supply different parts of the muscle, so that gradation of tension development by recruitment of active fibres is a possibility.

d. Flight Muscle

Before leaving the subject of neuromuscular transmission some account must be given of the activation of flight muscles in insects with so-called "asynchronous" or "indirect" flight musculature. The distinction between synchronous and asynchronous flight musculature was originally based on the observation that while the flight of insects like the locust, with wing-beat frequencies of about 20/s, could be adequately accounted for on the basis of known properties of the units of neuromuscular function (e.g. conduction velocity, latency of neuromuscular transmission, latency of contraction, contraction time and relaxation time), problems arose with certain insects for which wing-beat frequencies in excess of 1000/s have been recorded. On the basis of the anatomy of the pterothorax it was clear that each wing-beat cycle involved the contraction of antagonistic muscle pairs, acting as levators and depressors. With a wing-beat frequency of 1000/s, the time for a complete cycle would be 1 ms, and for the contraction and relaxation of one of the pair of antagonists 0.5 ms. When it is considered that a limiting factor for one of the fastest links in the chain of events, namely conduction, is the absolute refractory period of impulse conduction which exceeds 1 ms, it is clear that impulses could not be fired at the requisite frequencies in motor nerves to the flight muscle, much less could the activation of normal contractile machinery be accomplished in the requisite time. This startling anomaly prompted a thorough investigation

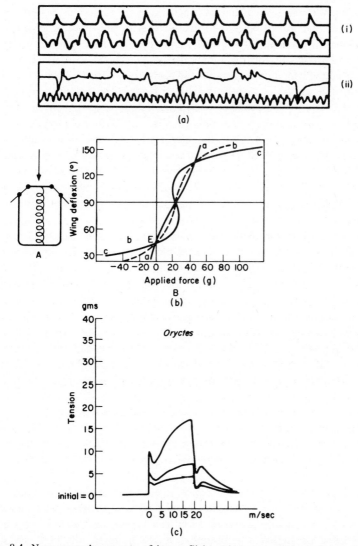

Fig. 8.4. Neuromuscular aspects of insect flight. (a) A comparison between electrical (upper traces) and mechanical (lower traces) records from the thorax of (i) a butterfly (synchronous), and (ii) a blowfly (asynchronous), during flight (Pringle, 1965). (b) A. Simplified model of an insect thorax: pressure in the direction of the arrow produces a click action only if there is an inward force from the sides of the model. B. Relationship in the model between vertically applied force and wing deflection: curve a, with no lateral stiffness; curves b and c, with increased lateral stiffness; E, equilibrium position of model as shown in A (Pringle, 1965). (c) Effect of small sudden changes of length on the tension developed by beetle fibrillar muscle. While developing a steady tension of 20 g, the muscles are quick-stretched and quick-released by various amounts at 0 and at 20 ms (Pringle, 1965; from Boettiger, unpublished).

by Pringle, who discovered that contractions of the flight musculature in insects of this kind did not bear a one-to-one relation to the firing of impulses in the motor nerve. Figure 8.4(a) shows a simultaneous comparison of electrical and mechanical events in the flight musculature of insects with synchronous and asynchronous flight musculature. In the butterfly (synchronous), contraction of flight muscles can be seen to be associated regularly with depolarizations of the muscle membrane, indicating a one-to-one relationship of contraction to the arrival of motor impulses, at a rate of about 50/s. In the blowfly, wing-beat frequency is about 160/s, but the rather irregular electrical events bear no relation to the mechanical events, and occur at a very much lower frequency. It appears that the asynchronous muscle is capable of going through a succession of contraction cycles under the influence of a single motor impulse, suggesting that the coupling between electrical and mechanical events is quite different from what it is in normal muscle.

Further investigation of the mechanisms involved in the flight of such insects revealed the existence of a peculiar mechanical system in the pterothorax. The flight muscles do not engage directly on the base of the wing, as they do in insects with direct flight musculature. Instead they extend dorsoventrally and longitudinally across the rigid box of the pterothorax, their contractions causing deformations of the mechanical structure that are transmitted to the wings in such a way that activation of the dorsoventral muscles causes wing elevation (see Fig. 8.4(b)), while the longitudinal muscles serve to depress the wings. The nature of the articulation and of the mechanical system is such that a "click"-mechanism is involved; for example, when wings are in the depressed position, and a force is applied to mimic contraction of dorsoventral muscles, elevation of the wings meets with increasing resistance until the wings click over abruptly into the elevated position (see curve c of Fig. 8.4(b)). The degree to which this phenomenon is manifested depends on the lateral stiffness of the thorax (cf. curve a), which in many insects is reinforced by special pleurosternal muscles.

The development of a type of "click"-mechanism is not, in fact, confined to insects with indirect flight muscles, but occurs also in species like the locust whose wings are directly activated. In all cases it ensures that a large proportion of the force of muscular contraction is taken up by the elasticity of the antagonistic muscles and by the cuticular structure of the thorax and wing articulation, and thus becomes available for moving the wings in the opposite direction at the end of a wingstroke. The development of a special elastic protein, resilin, in association with the wings of insects is of special significance in relation to the efficient storage of kinetic energy developed during the wingstroke, accounting for nearly a third of the energy stored in the locust. It is, however, in relation to the flight of insects with indirect musculature that the development of "click"-mechanisms becomes of special significance; for when the muscles have contracted up to a certain point (the point of equilibrium in Fig. 8.4(b)), there will be a sudden release of tension as the wing clicks into its new position of stability. This release seems to inactivate the muscle, but at the same time the tension which is suddenly applied to its antagonist causes that to develop active tension. The over-all result is a rapid alternate contraction of the antagonistic muscle units, the system behaving as a resonant mechanical oscillator, with the rate of oscillation determined principally by the mechanical properties of the pterothoracic box. It appears that the function of motor impulses, which arrive at the rate of one for every 4-10 oscillations, is to maintain the muscle fibres in an active state, capable of responding to the application of force in the way described.

Detailed investigation of these asynchronous (or fibrillar) muscles at the level of the contractile machinery have been undertaken by Pringle and his collaborators, using the large flight muscles of tropical beetles and bugs. These muscles show a characteristic response to externally applied load differing from that of normal muscle. Generally when muscle is subjected to a decrease or an increase in length during tetanic contraction, there is a transient rise, or fall, in tension, and the muscle then settles to a steady level characteristic of the new length. With fibrillar muscle, on the other hand, the immediate transient effects are followed by a further rise, or fall, after a brief delay (see Fig. 8.4(c)). In other words, there is a marked influence of the change in length on the contractile machinery of fibrillar muscle, and this appears to constitute its peculiar property, on which the contraction of asynchronous flight muscle is based. Further work has shown that the capacity for oscillation of the type described is a property of the contractile machinery as such, since muscle fibres from which most of the enzymatic and associated machinery has been removed by glycerination are capable of entering upon activity of this kind in the presence of ATP.

3. Synaptic Transmission

The unit of integrative activity in insects, as in other animals, is the synapse, and the question arises whether peculiarities of insect organization, including the limitation set by small size and the correspondingly small number of nervous elements at their disposal, might be reflected in corresponding peculiarities at the level of synaptic transmission. Before this problem can be considered it will be necessary to give some account of the extremely complex submicroscopic structure of the central nervous system of insects, which has been the subject of intensive investigation during recent years.

The neural elements of which the central nervous system is made up are basically of three kinds:

(a) the axon terminations of sensory cells conveying the input of information which serves as the raw material of integrative activity. The bipolar (type i) or multipolar (type ii) cell bodies of these neurones are situated at the periphery, in close association with the sense organs which they innervate;

(b) the unipolar interneurones, confined entirely within the central nervous system, and mediating between the sensory input and the motor output; and

(c) the motoneurones, whose cell bodies are located in the central nervous system, with an axon emerging from it to supply the muscles and effector organs of the body.

Associated with these neural elements in the central nervous system are numerous glial cells, of which two general types can be recognized (three, if the perineurium is included):

(a) those associated with the cell bodies, or perikarya, of the central nervous system; and

(b) those associated with the neuropile, the region of the central nervous system where axon collaterals of motor and internuncial neurones meet and mingle with the terminal branches of sensory fibres.

The general disposition of these elements is shown in Fig. 8.5(a), which represents a cross-section of part of an insect ganglion, ensheathed by neurilemma and perineural layers. The perikarya, with their associated glial cells, are situated peripherally in the ganglion, with the neuropile central, the two separated by a layer of glial cell bodies. The corresponding submicroscopic structure, as revealed mainly by the electron microscope, is one of enormous complexity, but the main features are made clear in the simplified diagram of Fig. 8.5(b).

Beneath the neurilemma and perineurium, which have been dealt with in an earlier section, lies the outer layer of the ganglion, composed of the cell bodies of motor and internuncial neurones together with the proximal parts of axons originating from them, and their associated glial cells. The anatomical relationship between them affords some clue to the role of glial cells in this region, for the cell bodies of the neurones are deeply indented by cytoplasmic processes of the glial cells to form the so-called trophospongium. The tremendous increase in the area of contact between perikaryon and glial cell suggests that the arrangement mediates an interchange of material between the two types of cell, and lends support to the view that the glial cells are concerned with the transfer of nutrients from the haemolymph, or from the fat bodies which often ensheath central nervous ganglia, to the ganglion cells. The glial cells, like those of the perineurium, contain numerous minute granules which are believed to be composed of glycogen, and a transfer of glycogen and lipid material from glial cell to the neurone perikaryon is suggested by the work of Wigglesworth (1960). It seems that this special mechanism of cellular nutrition may be a reflection of the lack of haemolymph circulation within the tissues of the ganglion; nutritive and excretory exchanges would lack the advantage of bulk flow in a circulating medium, and would need to be subserved by special mechanisms of active cellular transfer in the body of the organ.

A second characteristic feature of the outer glial layer is the extensive system of extracellular lacunae which is interposed between the ensheathed cell bodies and axons (see Fig. 8.5(b)). These lacunae often appear empty in electron microscope preparations, but usually they contain an electron dense material which may be a mucopolysaccharide. There can be little doubt that this system of lacunae represents the anatomical counterpart of the fast-exchanging extracellular space which has been described in an earlier section of this chapter, and it is possible that the anionic groups of the mucopolysaccharide which it appears to contain may act as a cation reservoir and thus play a part in the maintenance of a high extracellular sodium concentration.

The last feature of the outer glial layer which deserves mention is the sheathing of major axons which occurs in this region, by envelopment of the nerve fibres in folds of the glial cells, to form the so-called mesaxon (see Fig. 8.5(b)). This presumably serves to insulate the axon from presynaptic influences in this outer region of the ganglion, so that it is not until it reaches the neuropile that synaptic interactions can take place.

Interspersed between the outer glial layer and the neuropile is the second layer of glial cells, from the surface of which cytoplasmic processes extend to accompany the axons as they enter the neuropile. This produces a highly complex interdigitating network of cellular extensions, and the prospect of elucidating functional pathways within this network are remote. All that can be said at present is that axon profiles in the neuropile are often separated by the interposition of glial processes, and it may be presumed that under these circumstances synaptic interactions are likely to be precluded. A proportion of profiles, however, can be seen to be closely juxtaposed without intervention of glial membranes (see Fig. 8.5(b)) and it is often possible to distinguish what, by analogy with vertebrate systems, may be presumed to be presynaptic fibres containing aggregations of synaptic vesicles, from postsynaptic fibres whose cytoplasm is empty of such. It is not thought that such juxtaposition necessarily implies functional interaction, but in many regions a clustering of vesicles is associated with a thickening of both presynaptic and postsynaptic membranes, and these differentiated parts are likely to constitute active sites. What can be said, therefore, in general terms, is that the neuropile represents a part of the central nervous system where the transmission of excitation between axons of different neurones can be accomplished, and that the pattern of transmission, which forms the basis of integrative function, will be governed in part by the degree of overlap between axons from different neurones, and in part by the pattern of interposition of glial processes; these may be visualized as dissecting the neuropile into functional pathways, so forming a substratum for integration.

These studies of the submicroscopic anatomy of the central nervous system of insects have indicated that the fundamental basis of transmission is likely to be similar to that which has been experimentally established for other animals; they would then involve the release of chemical transmitters from presynaptic fibres, and the diffusion of such transmitters across narrow synaptic clefts to receptor sites on the postsynaptic membrane, where they would exert an effect based on alterations in membrane permeability and hence of membrane polarization. But while the basic mechanism of synaptic transmission may be the same, differences in the pattern of organization of synaptic interaction must be presumed to reflect important differences in the details of integrative function. In vertebrates the excitatory and inhibitory postsynaptic potentials are produced at synapses which encrust the perikaryon and its dendrites, and which can

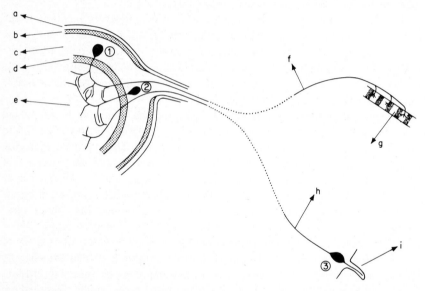

Fig. 8.5. The cellular structure of the central nervous system of insects. (a) Schematized plan of a section through an insect ganglion. a, neural sheath; b, perineurium; c, peripheral layer of perikarya with associated glial cells; d, inner layer of glial cells, associated with e, the neuropile; f, motor nerve; g, muscle fibre; h, sensory nerve; i, sense organ; 1, motoneurone cell body; 2, interneurone cell body; 3, sensory neurone cell body. (b) Diagram illustrating the disposition of cellular and extracellular components in an insect ganglion. Glial cytoplasm is indicated by light stippling, and extensive extracellular spaces by dark stippling. NL, neural sheath; PN, perineurium; OG, layer containing perikarya and associated glial cells, with an inner layer of glial cells associated with the neuropile, NP; ax_1, nerve fibre surrounded by concentric glial sheath; axon profiles 2-10 ensheathed by glial cells; other axon profiles closely apposed without glial separation, as shown at arrows. Large interconnecting extracellular spaces are present between the glial cell bodies and between their neuropile extensions, especially in the neighbourhood of tracheoles, tr. Note that the width of the region between perineurium and neuropile has been reduced for purposes of clarity, while the size of extracellular spaces between axon branches and glial processes in the neuropile has been exaggerated (Smith and Treherne, 1963).

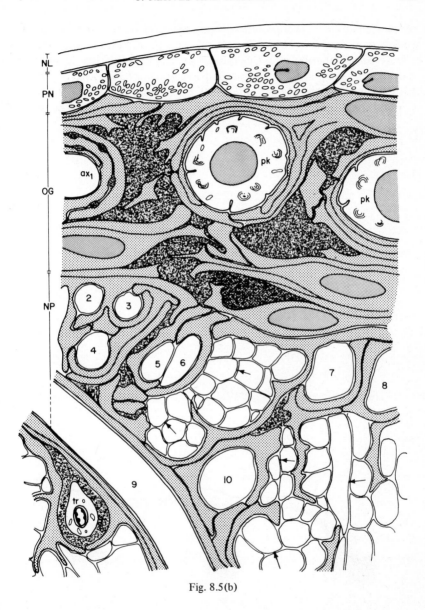

Fig. 8.5(b)

therefore be focussed on to a particular region of the neurone, the initial segment, where impulse initiation occurs. In insects one must imagine that the necessity to provide for the nutrition of the perikaryon by ensheathing glial cells would preclude the use of the cell body for synaptic transmission, so that this process has had to be relegated to sites of interaction on the axon itself or on axon collaterals. The profuse branching which characterizes the axons of insects would entail a wide spacing of sites of synaptic interaction, and this would militate against the use of a single focus for impulse initiation; it seems likely that impulses may, in fact, be generated at any one of a number of sites along the branched structure. In the absence of specific electrophysiological information one could speculate that such multiple impulse initiation may be of special

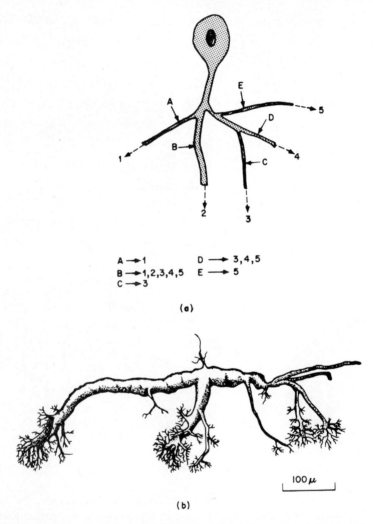

Fig. 8.6. Interneurones in insects. (a) Diagram of an interneurone, to illustrate how the output from the neurone might depend on the site of input. A-E, input sites; 1-5 outputs (Hughes, 1965). (b) The fourth giant internuncial from the central nervous system of *Gerris* (Hughes, 1965; from Guthrie).

significance in relation to the necessity for insects to effect nervous integration on the basis of a relatively small number of functional units. Figure 8.6(a) indicates the way in which different parts of a single neurone of the type described could function independently of each other, with impulses initiated at different points on the branched system depending on the nature of the input, and with the output governed in part by the site of impulse initiation, and in part by the ability of an impulse initiated in one region of the branched system to invade other regions, which could be a simple function of the relative diameter of branches. In this way the output from a neurone could take a number of different forms, and could be directed to a number of different parts of the nervous system, according to the pattern of input, and a single neurone could thus perform the function which on the vertebrate pattern would need a number of separate neurones. The bizarre configuration of certain insect interneurones (see Fig. 8.6(b)) suggests that the potentialities for this kind of effect may be considerable.

CHAPTER 9

SENSE ORGANS

The motor system and the central nervous system of insects have been shown to be fashioned from relatively small numbers of cellular elements, and the same applies in the sensory field. The density of sense organs on the surface of an insect's body is several orders of magnitude smaller than that associated with the body surface of a mammal; and where specialized receptor regions are concerned, as in the eye, the number of sense organs involved in insects is counted in thousands, in mammals in millions.

Certain other characteristics of the sensory physiology of insects appear to be referable to the possession of a rigid exoskeleton. This would constitute a formidable barrier to the more attenuated forms of environmental energy, and where such are concerned, as in olfaction and gustation, holes in the integument have had to be provided to allow the stimuli direct access to sensory elements. The monitoring of stresses set up in such a rigid exoskeleton has seemed to provide another special problem, solved by the development of sensillä which are sensitive to the shear forces arising as a result of mechanical deformation. Photoreception has remained relatively unaffected by the existence of a rigid exoskeleton, since the property of rigidity is not inconsistent with that of optical transparency.

a. Photoreceptors

Sensitivity to visible light is associated with two main receptor regions, the simple eyes, or ocelli, and the compound eyes. The structure and function of the compound eyes of insects have occupied the attention of insect physiologists for many decades, but despite the great amount of effort which has been devoted to the problem of insect vision, and despite the advances which the development of electron microscopy and the refinements of electrophysiological techniques have made possible, the physiology of insect vision remains poorly understood. The hypotheses of early physiologists concerning the mechanism of image formation have been put in question by recent experimental work, and lively controversies have developed on the subject of the system's optical properties, while the fundamentals of the photoreceptive process itself remain as yet completely unknown. It would clearly be inappropriate, in a book such as this, to attempt a detailed exposition of so controversial a field, and the present account will for this reason be a relatively superficial one.

(i) The Structure of the Compound Eye. The compound eyes of insects are made up of structural units called ommatidia, whose numbers range from a dozen or so to several thousand in different species. Each ommatidium may be considered to be made up of three functional parts (see Fig. 9.1(a) and (b)):

(a) the dioptric structures, which comprise a transparent part of the cuticle, known as the cornea, and a deeper-lying crystalline cone, through which light penetrates to the sense organs beneath;

(b) the photosensitive region of the ommatidium, known as the retinula, and composed usually of eight retinular cells. Each cell is a primary sensory neurone, continuous with a nerve fibre which passes, through the basement membrane that supports the sensory cell, to the central nervous system. The retinular cells are grouped around a central axis, and their photoreceptive regions, known as rhabdomeres, are centrally juxtaposed and sometimes fused to form an axial

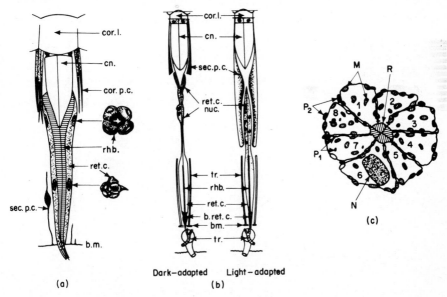

Fig. 9.1. The structure of the compound eye. (a) The ommatidium of a cockroach, and (b) of a moth in the dark-adapted and light-adapted states, showing migration of pigment, nuclei and cytoplasm. b.m., basement membrane; b.ret.c., basal retinular cell; cor.l., corneal lens; cor.p.c., corneal pigment cell; cn., crystalline cone; ret.c., retinular cell; ret.c.n., retinular cell nucleus; rhb., rhabdom; sec.p.c., secondary pigment cell; tr., trachea (Goldsmith, 1964, after Hesse; Umbach; and Day). (c) Cross-section of the retinular cells of the bee. M, mitochondrion; N, nucleus; P_1, pigment granules of retinular cells; P_2, pigment granules of pigment cells; R, rhabdom (Goldsmith, 1964).

rhabdom (see Fig. 9.1(c)). Closely packed arrays of microtubules are aligned at right-angles to the axis of each rhabdomere; and

(c) the pigment cells, which contain granules of red, yellow or brown pigment, and form a sheath round each ommatidium.

Two main kinds of ommatidia can be distinguished in different species of insect. In those that have the so-called "apposition" eyes, the rhabdom extends the full length of the ommatidium from basement membrane to crystalline cone, as in Fig. 9.1(a), and the distribution of pigment in the pigment cells is little affected by conditions of illumination. "Superposition" eyes occur typically in species that are nocturnal or crepuscular, and here the rhabdom is confined to the basal half of the ommatidium, and the distribution of pigment varies greatly depending on illumination (see Fig. 9.1(b)). In the dark-adapted eye the pigment granules aggregate distally in the pigment cells, while in the light-adapted condition they are more uniformly distributed between cornea and tip of rhabdom (for further details see (iv) below).

After passing through the basement membrane, the sensory axons enter the optic lobe of the brain, where they establish complex synaptic connections in a series of neuropiles designated as the first, second and third optic ganglia (Fig. 9.2). Two kinds of retinular axon can be distinguished, one of which has short

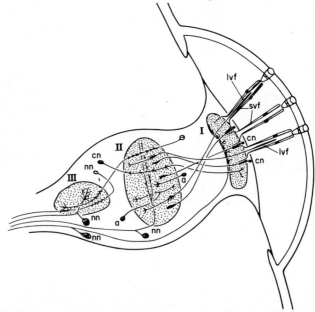

Fig. 9.2. Diagrammatic representation of neural connections in the eye and optic lobes of an insect. a, neurones associated with one synaptic region; cn, connecting neurones within the optic lobe; lvf, long visual fibre; nn, neurones connected to the central nervous system; svf, short visual fibre; I, II, III, first, second and third synaptic regions (Burtt and Catton, 1966).

fibres that terminate in the first optic ganglion, and establish synaptic contact with the lateral branches of multipolar interneurones. There appears to be much convergence in this region, each ganglion cell receiving input from sensory cells in neighbouring retinulae. Fibres from the second kind of retinular cell pass straight through the first ganglion and, accompanied by axons from the monopolar interneurones of that ganglion, they enter the second ganglion to establish synaptic contact with interneurones whose fibres pass to the central nervous system, as well as with interneurones that relay to the third ganglion. In addition to these centripetal pathways, there are centrifugal fibres originating in cell bodies deeper in the brain and passing outwards to branch profusely in the neuropile of the first ganglion.

(ii) *Pigments of the Compound Eye*. One of the most characteristic features of the insect eye is its heavy pigmentation, with pigment granules clustering not only in the special pigment cells, but also in the retinular cells themselves (see Fig. 9.1(c)). None of these pigments, however, appear to be directly involved in photoreception, since it has been shown that white-eyed mutants of several species of insect, which contain no trace of such pigments, are as sensitive to light as are the normal wild type. It seems likely that it is vitamin A and its aldehyde, retinene, both of which have been isolated from the heads of insects, that constitute the actual visual pigments in these as in other animals, the quantities involved being minute by comparison with those of the accessory pigments. These latter appear to function primarily as light shields, preventing stray light from moving obliquely through the eye, and they are of three main types, the ommochromes, the ommatins and the pterins (see Chapter 1). They have peak absorption in different regions of the spectrum, so that in combination they form an effective barrier to all but the longest wavelengths. The photolability of some of these pigments was originally thought to indicate a direct role in photoreception, but it now seems likely that if the phenomenon has biological significance it is in relation to the regulation of the amount of light that reaches the photoreceptive structures.

(iii) *Electrophysiology of the Compound Eye*. Because of the very small size of the retinular cells of the compound eye, it is only recently that progress has been made on the recording of unit events. Early investigators had to content themselves with investigations of the massed response to light of whole sheets of sensory cells, recorded in the form of a so-called "electroretinogram". Under these conditions a depolarizing response to light is obtained, which at low intensities is represented by a simple negative plateau, while at high intensities there is an initial fast phase which overshoots the negative plateau. The magnitude of the plateau is a close function of stimulation intensity. These potential changes appear now to be the expression of similar effects at the level of individual receptor cells, the responses of which are illustrated in Fig. 9.3. In view of the magnitude and positive sign of these potentials there can be little

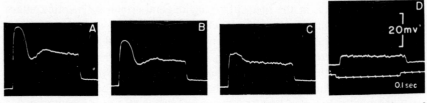

Fig. 9.3. Depolarizing potentials, presumably of a retinular cell, in the compound eye of a bee, recorded with intracellular electrodes during stimulation of the eye by light of different intensity, decreasing from A to D. Duration of stimulus 0.9 s (Goldsmith, 1964 from Naka and Eguchi).

doubt that they are developed across the membrane of the retinular cell, but the precise site of their origin has not yet been unequivocally established. Spike potentials have also been recorded by intracellular electrodes in the retinular cells, apparently initiated at the cell base or in the axon itself.

Measurements of the latency of spike responses recorded at various points in the optic pathway between the retina and the ventral nerve cord following stimulation by light suggest that, at moderate intensities, the excitation is relayed successively through each of the three optic lobes, but with stronger stimuli some short-circuiting of synapses in the second and third ganglia occurs, to give a reduced latency. Unit activity of elements in the second ganglion have been investigated, and a high proportion of cells were found to give bursts of spikes at light-on and light-off. Evidence was obtained of substantial convergence, some units responding to illumination of any point in the visual field.

(iv) Light and Dark Adaptation. In the compound eye, as in photoreceptors of other types, there is a decrease in the sensitivity to light during exposure to light (light adaptation), and a recovery during darkness following such exposure (dark adaptation). In insects with apposition eyes, where there is little movement of pigment, dark adaptation proceeds to completion within 20 min, the initial stages of recovery being very rapid. In insects with superposition eyes, the curve has an initial rapid phase, followed by a much slower phase of recovery, as illustrated in Fig. 9.4. The initial phase is similar to the recovery of sensitivity in apposition eyes, and may be attributed to some process of recovery in the photosensitive process whose precise nature has not yet been identified, but which may involve a resynthesis of visual pigment. The second, slower phase is closely associated with, and appears to be caused by, migration of the accessory pigments in the pigment cells (see inset of Fig. 9.4). This is confirmed by the fact that in a certain proportion of individuals there is a failure of such pigment migration, and where this occurs there is failure, too, of recovery beyond the level of the first part of the curve.

(v) The Physiology of Vision. While there has been considerable progress during recent years in the exploration of fine structure and of unit electrophysiological correlates of photoreception, the complexity of the architecture of

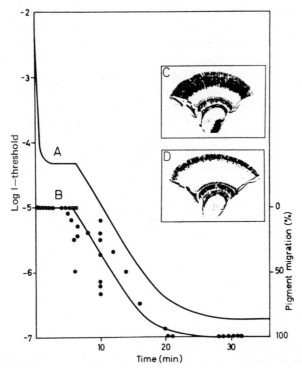

Fig. 9.4. Dark adaptation (A) and pigment migration (B) in *Cerapteryx graminis*. Curve A derives from measurement of 22 preparations; curve B shows the composite results obtained from 90 preparations representing average values of measurements in 2-5 specimens. Pigment migrations are given as a percentage of the distance between extreme light (inset C) and dark position (inset D) of the proximal border of the secondary pigment (Goldsmith, 1964 from Bernhard, Höglund and Ottoson).

the compound eye and its associated neuropiles has so far precluded a convincing interpretation of the visual process in terms of the properties of its constituent elements. Behavioural experiments with a wide variety of insects have established the existence of well-developed form discrimination (see Chapter 11), but the physiological basis of this capacity remains obscure. A general theory of insect vision was put forward early in the history of investigations by Exner, who proposed that each ommatidium would be sensitive only to light entering at a small angle to its axis, other rays being absorbed by the screening sleeves of pigment cells. There would therefore be little overlap between the visual fields of adjacent ommatidia, and the image formed on the retinular layer of sensory elements would be essentially a mosaic. The situation would be slightly different in dark-adapted superposition eyes where, to ensure increased sensitivity, a peripheral migration of the pigment sleeves would allow light coming through one ommatidium to affect the photoreceptive regions of adjacent ommatidia. Objections are currently being raised to this interpretation on the basis of recent experimental evidence, but no completely satisfactory alternative has so far been proposed. It therefore remains uncertain precisely what the nature is of the image which is projected on to the sensory elements, and what the mechanism is of its projection, nor is it known to what extent the receptive process may be under the control of centrifugal and lateral nervous influences analogous to those involved in vertebrate vision.

The situation in regard to another important aspect of the physiology of vision, namely colour vision, is rather more satisfactory. The existence of colour vision in insects was established a long time ago on behavioural criteria, and its electrophysiological basis has now been uncovered. It has been possible to change the spectral sensitivity curve of certain insects by adapting the eye to coloured light, and in this way the existence of two distinct visual pigments in the eye has been demonstrated, one with a peak at 340 mμ, and one with a peak at 540 mμ, thought to be located in different receptor cells. In flies it has been possible to investigate the spectral sensitivity of single sense cells using intracellular electrodes; all the cells tested showed a peak at 350 mμ, but while some had a second maximum at 450 mμ, in others the secondary peaks fell at about 480 and 520 mμ (see Fig. 9.5). The existence of different sensory neurones with different spectral characteristics of this sort is adequate to account for the existence of colour vision, as established on the basis of behavioural experiments.

(vi) Ocelli. Simple eyes occur in the adults of most winged insects, situated on

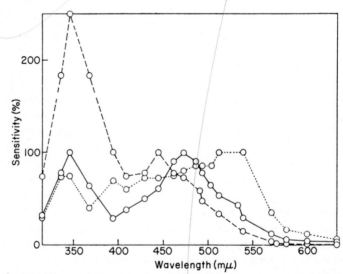

Fig. 9.5. Spectral sensitivities of three individual retinular cells of the compound eye of a fly, measured with an intracellular microelectrode (Goldsmith, 1964 from Burkhardt).

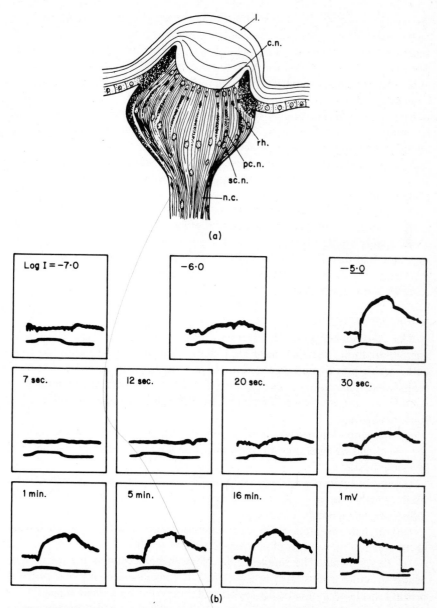

(a)

(b)

Fig. 9.6. Simple eyes of insects. (a) Section through the ocellus of *Arthrophora spurmaria*. c.n., nucleus of cornea-forming cell; l., lens; n.c., ocellar nerve; pc.n., nucleus of pigment cell; rh., rhabdom; sc.n., nucleus of retinula (Imms, 1948). (b) Potentials recorded from the ocellus of the cockroach in response to light of increasing intensity (top row) and in response to light of log intensity = −5 at different intervals during dark adaptation following light adaptation for 1 min at 12,000 ft candles. The duration of the test stimulus is shown on the lower trace of each record (Ruck, 1958).

the frontal region of the head between the compound eyes. They consist of an aggregation of light-sensitive cells, closely resembling the retinular cells of the compound eye, covered by a simple "corneal" lens of transparent cuticle (Fig. 9.6(a)). The short neurones converge to synapse below the retina with dendrites of a relatively small number of second-order neurones that extend from the brain in the form of an ocellar "nerve".

Stimulation of the ocellus with bright light causes depolarization of the retinular cells, the magnitude of the response being related to stimulation intensity and to the degree of dark-adaptation (Fig. 9.6(b)). The sense cell depolarizations in turn produce hyperpolarizing postsynaptic potentials which serve to silence the spontaneous discharge of the ocellar fibres during the period of illumination.

The role of ocelli has not yet been elucidated satisfactorily; in view of the great convergence there can be no question of image resolution, nor do the ocelli alone appear to be capable of mediating phototactic responses. It has been suggested that they may exert an indirect effect on visual and other reactions, by increasing the level of excitation of corresponding nerve centres, but the evidence is not convincing.

b. Mechanoreceptors

The mechanoreceptors of insects are characterized by a great diversity of anatomical form, and they have been classified in different ways by different authors. For present purposes some attempt must be made to reduce to simple terms what is in reality a most complex situation, and this has been done by considering insect mechanoreceptors as belonging to one of three main functional groups:

(i) mechanoreceptors which effectively extend the zone of contact between the insect and its environment, projecting from the insect's surface as some kind of tactile bristle;

(ii) mechanoreceptors which monitor the stresses set up in the exoskeleton itself, whether as a result of outside influences, like gravity, or of inside influences, like muscular contraction;

(iii) mechanoreceptors which register tensions within the body, arising usually as the result of movements of one part relative to another.

(i) Tactile Setae. The simplest type of mechanoreceptor is the tactile seta, whose structure is illustrated in Fig. 9.7(a). It consists essentially of a cuticular bristle articulated in a cuticular socket and innervated by a single bipolar sense cell. Associated with the sensillum are the trichogen and tormogen cells, responsible for the formation of the cuticular parts of the sensillum, and an ensheathing neurilemmal cell. The distal process of the neurone ends as a terminal filament, or scolopid body, at the margin of the socket, while its proximal axon connects to the central nervous system. Displacement of the setae

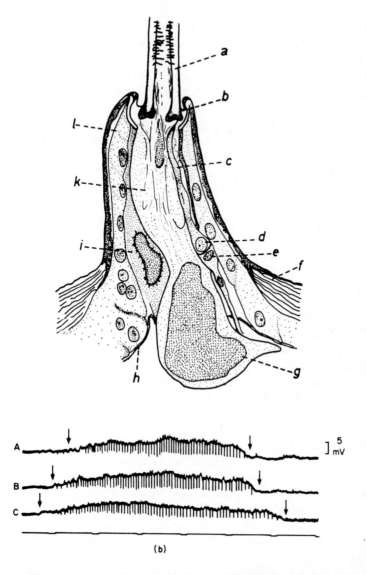

(b)

Fig. 9.7. Tactile setae. (a) Section through the base of a tactile bristle from the larva of a butterfly. a, base of hair; b, articular membrane; c, scolopoid body; d, sense cell; e, neurilemma cell; f, cuticle; g, trichogen cell; h, basement membrane; i, tormogen cell; k, vacuole; 1, epidermis (Schwartzkopff, 1964 after Hsü). (b) Electrical responses of a mechanosensory hair from the blowfly to mechanical deformations of progressively smaller intensity (A, B and C). Arrows mark the beginning and end of stimulation; note that differences in the amount of deformation are reflected in the size of the generator potential, the size of the action potentials and the frequency of firing. Time marks at 0.2 s intervals (Dethier, 1963; courtesy, M. L. Wolbarsht).

causes mechanical deformation of the sensory process, and so sets up a graded generator potential which in turn initiates a discharge of spikes in the axon. The frequency and size of sensory nerve impulses are related directly to the magnitude of the generator potential, as indicated in Fig. 9.7(b).

While the basic anatomical structure of tactile setae is fairly constant, the precise nature of their response to stimulation shows considerable variation. At one extreme are the delicate tactile hairs which are so sensitive to stimulation that a gentle puff of air is sufficient to initiate discharge. These are generally of the "phasic" type, characterized by rapid adaptation; they give short, high-frequency bursts of impulses at the moment of deflection, but remain silent under conditions of constant deformation. At the other extreme are the stout bristles, whose displacement requires considerably greater force. These show a high rate of discharge during initial phases of stimulation, but the frequency drops to lower, steady levels under constant deformation. The rate of discharge of these "tonic" receptors often depends on the direction, as well as on the extent, of displacement.

(ii) Stress Receptors. Cuticular stresses are registered by a special type of sense organ known as a campaniform sensillum, which consists of a canal in the cuticle, covered by an elliptical, domed cap, and innervated by a single bipolar neurone, as illustrated in Fig. 9.8(a). The mode of action of these sense organs was established by the work of Pringle, who showed that tonic discharges could be initiated in the sensory nerve by suitable mechanical deformation of the cuticle, the discharge frequency depending on the extent of deformation (see Fig. 9.8(b)). Such deformations had the effect of increasing or decreasing the convexity of the dome, and hence of increasing or decreasing tensions acting on the terminal filament attached to it. Consideration of the geometry of the physical system suggests that the sensillum responds to the compression component of the shear force developed by mechanical deformation, and that the orientation of the sensillum will determine the orientation of shear force to which it is sensitive. This view receives support from a consideration of the distribution of sensilla (see Fig. 9.8(c)); the sense organs tend to be grouped into what would appear to serve as functional units, each sensitive to a different component of shear force, and hence to a particular form of mechanical deformation.

(iii) Stretch Receptors. The multipolar (type ii) neurones, which are widely

---

Fig. 9.8. Campaniform sensilla. (a) Section through a campaniform sensillum from the cockroach. a, dome-shaped plaque; b, scolopale; c, vacuole in membrane forming cell; d, accessory cell; e, sense cell; f, neurilemma cell (Wigglesworth, 1965 after Hsü). (b) The pattern of discharge from campaniform sensilla on the maxillary palps of the cockroach stimulated by stronger (curve a) or weaker (curve b) stresses in the cuticle, produced by bending (redrawn from Pringle, 1938). (c) Details of the orientation of the six groups of campaniform sensilla on the third leg of the cockroach (Pringle, 1938).

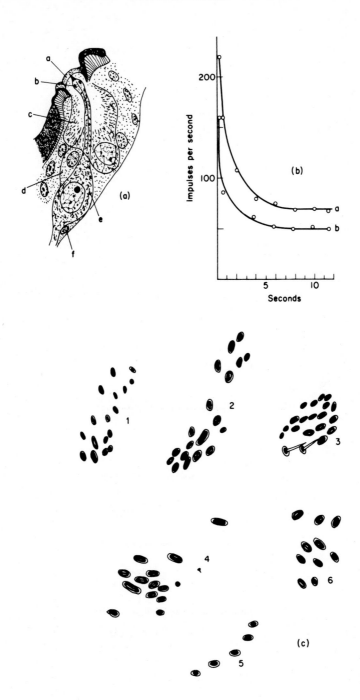

(a)

(b)

(c)

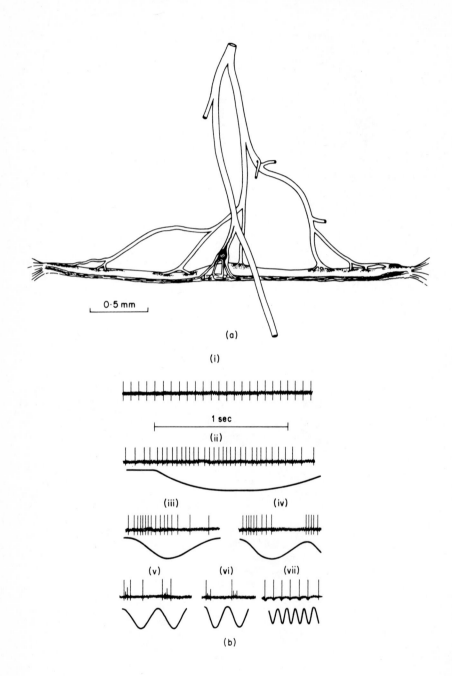

0·5 mm

(a)

(i)

1 sec

(ii)

(iii)   (iv)

(v)   (vi)   (vii)

(b)

distributed among the internal organs of insects, have been intensively studied by Finlayson and Lowenstein, who were the first to establish the function of these structures in insects as stretch receptors. A number of different types have been described, some associated with strands of connective tissue, some with muscle, and some with specialized muscle fibres to form sense organs reminiscent of the vertebrate muscle spindle (Fig. 9.9(a)).

In the complete absence of stretch there is no discharge from these sense organs, but under slight stretch, which is the normal condition in the body, there is a low basal rate of non-adapting discharge at 5-10 impulses/s (see trace (i) of Fig. 9.9(b)), which continues steady for hours on end. Increasing tension leads to an increase in discharge frequency, which is maintained at the higher level, with frequency closely proportional to the degree of stretch. The receptors thus conform to the classical picture of the tonic receptor, but they are in fact capable of performing a dual function. Experiments involving alternate stretch and relaxation of the sense organs showed that discharge frequency under these conditions was a function not only of the degree of displacement, but also of the velocity of displacement (Fig. 9.9(b), traces (ii)-(vii)). With increasing frequency of displacement the tonic element of the discharge tends to drop out, and the fibre remains silent during periods of maximum displacement as well as during relaxation (traces (vi)-(vii)), possibly as a result of postexcitatory inhibition; at these frequencies (above 5 cyc/s) the receptors therefore behave in a phasic manner, monitoring frequency of stretch rather than degree of stretch.

Another type of mechanoreceptor that may respond to stretch is the chordotonal organ, consisting of one or more type (i) neurones, associated with two companion cells and slung between one point of the body wall and another. These sense organs appear generally to subserve a proprioceptive function, but in many cases they are associated with specialized auditory membranes for the reception of airborne vibrations (see Chapter 10 for further details).

This brief account of the mechanoreceptors of insects has done less than justice to the amazing variety of structure and function which is exhibited by this type of sense organ within the group. It must be emphasized, too, that the simple classification which has been adopted cuts across the physiological distinction which can usually be drawn between exteroceptors and propriocep-tors. Examples from any of the three categories may function as either,

---

Fig. 9.9. Stretch receptors. (a) Diagram of the structure and innervation of a stretch receptor from the pupa of a moth *(Antheraea)*, showing the modified muscle fibres (upper) innervated by branches from a motor nerve, and the multipolar neurone, whose dendritic processes run alongside the muscle fibre on its lower surface (Finlayson and Lowenstein, 1958). (b) The response of a stretch receptor from the pupa of a moth to phasic stimulation; the upper trace in each record shows the neurone action potentials, the lower monitors mechanical displacement: (i) resting discharge; (ii) stretch at 0.5 cyc/s; (iii) at 1.5 cyc/s; (iv) at 2.0 cyc/s; (v) at 4.0 cyc/s; (vi) at 5.0 cyc/s; (vii) at 15.5 cyc/s; (schematized from oscilloscope photographs of Lowenstein and Finlayson, 1960).

depending on their location and the extent to which they are functionally coupled with similar units. For instance, while tactile setae normally could be considered as exteroceptive, in many insects they are grouped to form hair-plates near points of articulation of the exoskeleton, and under these circumstances they would function as proprioceptors, providing information about the relative position of parts of the body. Similarly, the stress receptors of the cuticle, while normally fulfilling a proprioceptive function, would register stresses imposed on the insect's body by gravity or the pressure of wind, and could in this sense be considered as exteroceptors; and chordotonal sensilla, whose primary function could be considered as proprioceptive, are often associated with accessory structures, as at the tympanic membrane or in the Johnston's organ of the antenna, and there serve a function of exteroception in relation to airborne vibrations. It is the very diversity of form and function of insect mechano-receptors that has necessitated a superficial treatment and so created a false impression of simplicity.

c. Chemoreceptors

In insects, as in other terrestrial animals, a distinction is often made between two types of chemoreceptor; the one receptive to vapours at relatively low concentration, normally referred to as olfactory; and the other mediating a response to substances in solution at relatively high concentration, usually called a gustatory, or contact, chemoreceptor. In insects the two types have one thing in common, in that the receptive surfaces of both are bare to the influence of the environment through some kind of aperture in the general cuticular investment. Figure 9.10(a) shows an olfactory sensillum from the antenna of a grasshopper, innervated by a considerable number of bipolar neurones, whose distal processes project up into the perforated cuticular peg to terminate as clusters of microvillar dendrites in the aperture of perforations. In the much investigated contact chemoreceptors, the number of neurones is very much smaller, and their terminal processes extend into the shaft of the chemosensory hair to end just below a terminal pore (see Fig. 9.10(b)).

Early studies of the physiology of chemoreception were based largely on the stereotyped responses which in many species of insect can be elicited by suitable stimulation of chemoreceptors. For instance, if the tarsal sensilla of a hungry blowfly are brought into contact with a solution of sucrose, the proboscis of the fly is usually lowered in preparation for the act of feeding. By determining the threshold for proboscis extension with different materials which evoke a positive response, or alternatively by determining the rejection thresholds for substances which cause rejection when presented together with a substance that would otherwise elicit a positive response, a lot of useful information has been obtained concerning the physiology of the receptor process. It has been shown, for instance, that the rejection threshold for homologous series of alcohols and

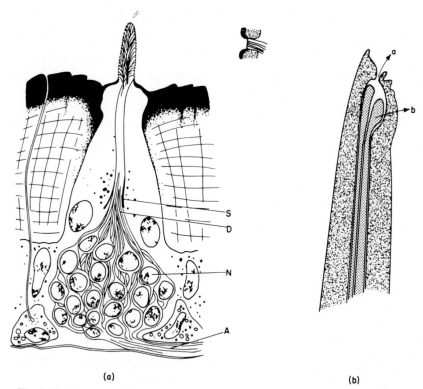

(a)   (b)

Fig. 9.10. Insect chemoreceptors. (a) Section through an olfactory sensillum from the antenna of a grasshopper, showing dendrites (D) from a large number of bipolar neurones (N) enclosed for part of their length in a scolopoid sheath (S) extending into the peg, and terminating at the pegwall perforations. A, axons. The inset shows a section through one of the perforations with dendritic microvillae in the aperture (Dethier, 1963). (b) Section through the tip of a tarsal chemoreceptor of the stable fly, showing the pore (a) at the end of the sensillum, below which the neurone dendrites (b) terminate (drawn from electron micrograph of Hodgson, 1964, courtesy Adams).

glycerols decreases with increasing length of the carbon chain, and that a discontinuity occurs in this relation at the point where oil/water partition coefficients show an abrupt change, suggesting that the effectiveness of compounds in causing rejection is in part associated with lipid solubility.

The main objection to the use of behavioural criteria for investigations of sensory physiology is that what is being tested is not the sensitivity of the sense organ, but the reactivity of the whole of the neuronal network interposed between the sensory and the motor fields. Critical exploration of chemosensory physiology had, therefore, to await the development of electrophysiological techniques sufficiently refined to detect the activity of the sensory neurones themselves. Olfactory receptors have proved rather intractable in this respect, and most investigators have here had to rely on a study of summed responses in large numbers of antennal receptors to provide a measure of activity. Such "electroantennograms" have been shown to differ in form depending on the nature of the stimulating odour (see Fig. 9.11(a)), particularly characteristic patterns being obtained with the antennae of male silkworms when they are stimulated by the female sex-attractant. The very large discrepancy between the concentration of attractant necessary to elicit a distinctive electrophysiological response and that required to initiate the corresponding behavioural reaction suggests, however, that the approach is too crude to be of use for investigations of the neurophysiological basis of the response.

Electrophysiological investigations of gustatory sense organs have met with considerably more success. Here the number of sense cells involved in a single sensillum is small, and the response of different neurones can often be distinguished on the basis of spike height. By the ingenious use of capillary electrodes placed over the tip of the sensillum, providing at the same time a means of monitoring the electrical potentials developed as a result of stimulation and a means of applying the gustatory stimulus, the appropriate substances being incorporated in the solution of the capillary electrode, it has been possible to investigate the physiology of chemoreception at the level of the sensory neurone. In the labellar chemoreceptors of the blowfly two types of afferent potential, large and small, can be distinguished (Fig. 9.11(b)). These have been shown to be evoked in response to the application of inorganic salts and of sugars respectively, and the frequency of impulse discharge has been found to be linearly related to concentration over the range of 0.03-1.00 M.

The specificity of the response to sugar was investigated by testing a variety of different types of carbohydrate. One of the most important properties of active substances was found to be the presence of an α-D-glucopyranoside linkage, as in sucrose and maltose, a finding which is fully substantiated by behavioural studies. It is postulated that the terminal surface of the carbo-hydrate receptor incorporates a highly specific receptor site, with which only sugars of the right molecular configuration can react; and that the interaction between receptor surface and stimulating molecule in some way causes a depolarization of the membrane to produce a generator potential, which in turn initiates a discharge of action potentials in the afferent nerve. It has recently proved possible to monitor the graded generator potential itself (see Fig. 9.11(c)), by recording from the fractured base of the chemosensory hair.

A third chemoreceptor has been identified in the labellar hair of the blowfly, responding to the application of pure water by low amplitude impulses detectable through the side wall of the sensillum. The discharge is inhibited by sucrose, and provides a neurophysiological basis for the ability of blowflies to distinguish water from other acceptable substances. It should also be mentioned that a mechanoreceptive neurone is associated with the articulated socket of the chemosensory hair, giving the single sensillum a dual function.

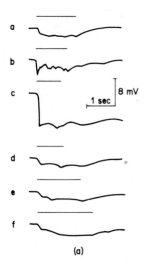

(a)

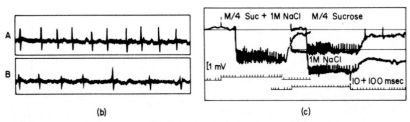

(b)   (c)

Fig. 9.11. Aspects of the electrophysiology of chemoreception. (a) Electroantennogram from the isolated antenna of a female moth subjected to stimulation by currents of air carrying different vapours. a, unpurified air; b, air plus wintergreen-oil vapour; c, air plus clove-oil vapour; d, unpurified air; e, air plus female sex-attractant of same species; f, air plus sex attractant of different species. Note the absence of response in the female antenna to sex-attractant which in a male antenna would give a response similar to trace c. The duration of air currents is marked by horizontal bars above the EAG (Schneider, 1962). (b) Spike potentials recorded by capillary electrodes on the labellar chemosensory hairs of the blowfly. A, the large fibre response to stimulation with 0.5 M NaCl; B, large and small fibre response to stimulation by a mixture of 0.25 M sucrose and 0.1 M NaCl (Hodgson, 1964). (c) Generator potentials and spike potentials from labellar hair of blowfly. Single stimuli applied to obtain record on the right. The left record made following application of a mixture of salt and sugar, showing partial summation of generator potentials (Hodgson, 1964, courtesy Marita).

d. Humidity Receptors

Brief mention must be made of the ability of many insects to respond to humidity, to water as a vapour rather than as a solvent. The sense organs involved have been identified on behavioural criteria in a number of insects, but nothing is known of the electrophysiological basis of humidity reception. In some cases it has been possible to establish that the response is to evaporating power rather than to water vapour, in others relative humidity appears to be the effective stimulus. Precisely what it is that mediates afferent discharge is, however, still unknown; it might be water vapour acting as an olfactory stimulus; it might be changes in the physical conformation of receptor structures based on cuticular hygroscopy; or it might be changes in osmotic conditions, or in temperature, associated with the evaporation of water from receptor surfaces. Until electrophysiological methods can be refined to the point where changes in the electrical potential of receptor membranes can be detected, there seems little hope for progress in this field.

e. Temperature Sensitivity

Many insects are known to respond to temperature, and the antennae are often involved in such responses. It is possible that the reaction may be mediated by receptors sensitive to other modalities, whose activity may be modulated by changes in temperature, as the activity of the chemosensory neurones is known to be. The high degree of discrimination shown by certain species makes it likely, however, that specific temperature receptors exist, though their identity has not yet been unequivocally established.

CHAPTER 10

INTEGRATIVE ASPECTS OF NERVOUS FUNCTION

While important advances have been made during recent years in the elucidation of unit function, progress in the field of nervous integration has been slow. One might imagine that insects would constitute particularly favourable material for the investigation of the physiology of integration, in view of the relatively small number of elements which are involved both on the sensory and the motor side. It seems, however, that the advantages conferred by neural parsimony are to a large extent offset by technical difficulties associated with the special architecture of the central nervous system in insects. In the first place, there is no convenient separation of afferent from efferent routes in insect nerves; the nerves are mixed from their point of origin in the central nervous system, carrying both sensory and motor fibres, so that neither in the stimulation of nerves, nor in recording from them, is it possible to distinguish the two parts of the reflex arc. The second difficulty arises from the nature of the neuropile in insects. The large neurone cell bodies, from which intracellular recording would be technically possible, are situated at points remote from the interplay of synaptic effects, and provide correspondingly little evidence of the nature of such effects. To investigate the electrical activity of the neuropile, recourse must be had to extracellular recording, and the results have so far been disappointing; for while it is possible to pick up active units by this means, such units appear to be sparsely distributed, and separated by large tracts of electrically silent tissue. The active units can often be shown to come under the influence of sensory input, as illustrated in Fig. 10.1, where the response of a brain unit to flashes of light is recorded. But the lack of variety and discrimination in the pattern of response of such units has been a cause for concern among investigators. The severe restraint under which the insects have to be placed in order to permit recording may be in some part to blame for the apparent lack of meaningful response, for the fact that, as one research worker puts it, "almost all the recordings can be interpreted as alarm responses dulled by repetition". But apart from this, the prospects for an elucidation of the neurophysiological basis of integration will remain poor so long as the activity of a high proportion of neuropile units defies detection with present instrumenta-

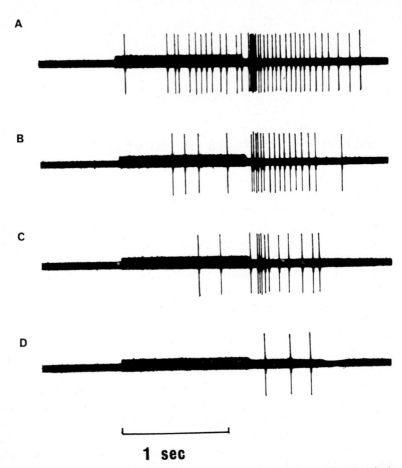

1 sec

Fig. 10.1. The response of a brain unit to successive presentations of identical flashes of light presented at 5-s intervals. The long latency at "on", together with the pause after the initial spike in the first line, suggests that the "on" response has been suppressed by inhibition. The "off" response shows progressive habituation (Horridge *et al.*, 1965).

tion, since it is presumably the smaller, silent elements that serve as the basis of integrative activity.

One other factor militates against progress in the field of central nervous function, and that is the lack of accurate maps of the general anatomy of the central nervous system in insects; only a few species have been carefully investigated, and these not the ones that are most suitable as experimental material. It would seem that until this deficiency has been made good, and until refinements of neurophysiological technique open the way to investigation of smaller neuropile elements, the finer details of integrative function will remain beyond our reach. The best that can be done for the present will be to establish the relation between input to the central nervous system and output from it. Both can often be precisely defined in terms of unit activity, and the relation between them can be determined in quantitative terms; how, precisely, this relation is established within the integrative networks of the neuropile will have to remain a problem for the future.

The pattern of input from different types of sense organ has been briefly outlined in the previous chapter; in the first part of this one, a few cases will be discussed where attempts have been made to relate changes in input, as defined in electrophysiological terms, to changes in output, as defined in broader behavioural terms. In the second part the approach will be made from the motor side of the reflex pathway, and the pattern of activity in the final common path, defined in electrophysiological terms, will be related to changes in input which can sometimes be defined in quantitative terms, sometimes only in general terms. In the last part consideration will be given to the effect on output of operative or electrical interference with nerve centres remote from the final common path. Here the nature of the interference can only be defined in gross anatomical terms, and interpretation becomes correspondingly uncertain.

1. The Neurophysiology of Escape Reactions

a. The Startle Response of the Cockroach

One of the most striking characteristics of the central nervous system of insects, as of many other invertebrates, is the so-called giant fibre system. This generally mediates escape reactions of some kind, as in the startle response of the cockroach, where the speed with which the reaction can be performed may have dramatic adaptive significance, making the difference between the life and death of the individual concerned. The response latency of such reactions has in many species been greatly reduced by an increase in the diameter of certain internuncial nerve fibres (see Fig. 10.2(a)), which has resulted in a corresponding increase in the conduction velocity. Because of their large size, the activity of these giant interneurones is readily monitored, and most of the reflex components are thus open to electrophysiological investigation. The startle response of the cockroach has been particularly carefully studied, and details of the reflex pathway between the sensory field, as represented by mechano-receptors on the anal cerci and the motor field, as represented by the musculature of the metathoracic leg, are illustrated in Fig. 10.2(b).

Stimulation of the delicate tactile sensilla by a puff of air leads to a massive discharge of sensory neurones and provides an afferent input to the giant internuncials in the last abdominal ganglion. The size of the resulting excitatory postsynaptic potential developed across the giant fibre membrane, increases with increasing intensity of stimulation, by spatial and temporal summation,

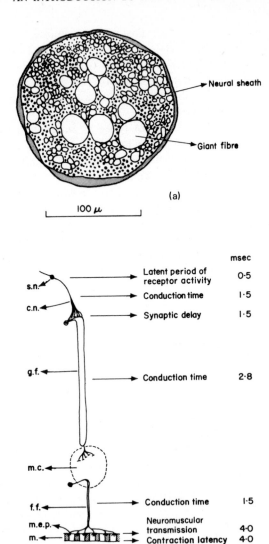

Fig. 10.2. Neuromuscular basis of the startle response of the cockroach. (a) Giant fibres in a cross-section of the ventral nerve cord of the cockroach (drawn from photomicrograph in Huber, 1965, from Roeder). (b) Diagram of the neuromuscular elements involved in the startle response of the cockroach, with the duration of unit events listed at the right of the figure. The linear scale has been greatly distorted for ease of representation. c.n., cercal nerve; f.f., fast fibre; g.f., giant fibre; m, muscle; m.e.p., motor end plate; m.c., motor centre; s.n., sensory neurone (from Roeder, 1963).

until it is sufficient to trigger a discharge of action potentials in the giant fibres. The impulses sweep up through a succession of abdominal ganglia to the metathoracic ganglion, where the giant fibres divide to form a tangle of smaller branches embedded in the neuropile. They transmit ultimately to the fast motoneurones of the metathoracic leg, whose discharge causes contraction of the leg muscles, and so initiates the escape reaction.

Most of the unit events in the reflex pathway have been accurately timed, and estimates of duration have been included in Fig. 10.2(b); the sum for known unit elements is about 16 ms. There is a substantial difference between this value and experimentally determined latencies for the reaction as a whole, which average 54 ms, with a minimum of 26 ms. The difference must clearly be attributed to the time taken for transmission through what may be called the "motor centre" of the metathoracic ganglion. What is involved in the motor effect is not simply the contraction of a muscle, but rather the co-ordinated movement of a limb, and the substantial time that intervenes between the arrival of giant fibre impulses and the initiation of activity in the final common paths suggests that a neuronal network of considerable complexity may be interposed at this point. This is borne out by the very labile and unpredictable nature of the response at this level of the reflex arc, which has so far precluded detailed electrophysiological analysis. Transmission through the metathoracic ganglion often appears to block irreversibly under experimental conditions, and under natural conditions it is this point that appears to come under inhibitory influences, which may cause the behaviour pattern to extinguish completely after only a few trials. This is an example of the general phenomenon of habituation, which will be discussed further in Chapter 11.

b. Predator Evasion by Noctuid Moths

The startle response of the cockroach is essentially an all-or-nothing type of response, involving a minimum of discrimination at the level of input; the motor pattern does not appear to be directed in relation to the stimulus, nor guided by its form. Either the sensory discharge is sufficient to initiate activity in the giant fibres or it is not; if it is, then the behaviour pattern may be triggered off, or it may not, depending on the state of the motor centre, which is the point at which discrimination appears to occur. By contrast, the evasive reactions of certain noctuid moths during pursuit by bats is closely related to the nature and the source of stimulation, reflecting a much greater content of information at the level of the sensory input. In this case, however, no giant internuncials are involved, and the input is lost in the silent regions of the neuropile as soon as it enters the central nervous system. The relation between input and output can therefore only be gauged by observation of behavioural performance.

The sense organs involved in predator evasion are the paired tympanic organs, each consisting essentially of a thin membrane stretched on a cuticular frame.

Movements of the membrane induced by airborne vibrations cause a discharge in two sensory neurones associated with the tympanum. The sensory physiology of this "ear" has been investigated in detail by Roeder, whose results show that it is sensitive to a very wide range of frequencies, with a sensitivity maximum at about 60 Kc/s, a frequency which is well represented in the ultrasonic, echo-locating cries of insectivorous bats. The discharge frequency in the tympanic nerve is a linear function of intensity over a range of about 40 decibels above which the receptors saturate.

The sensory basis of predator evasion is thus of the greatest simplicity, but in spite of this the information content of the sensory input is substantial, as illustrated by records of tympanic discharge evoked by the echo-locating cries of a hunting bat. The cries consist of short bursts of high frequency sound produced at a rate of about 10/s during cruising, with the pattern changing to a rising crescendo or "buzz" of pulses delivered at a higher rate, as the bat closes in upon its prey. In Fig. 10.3(a) the discharges of left and right tympanic nerves of a restrained moth are recorded in the presence of a hunting bat. Record A shows the response to an approach by a cruising bat, signalled at first by a slight discharge from only one tympanic membrane, thus providing information concerning the direction of approach. The second cry is picked up by both organs, the higher intensity of discharge indicating that the bat is approaching; information about direction is still available on the basis of differences in discharge frequency. The third response is similar, but in the fourth the pattern of discharge on the two sides is virtually identical, indicating that at this stage the bat is directly overhead. In record B the response to the closing "buzz" of a hunting bat is shown, as registered mainly by one of the tympanic organs, while in record C the "buzz" is registered by both.

These results show the mechanism by which the discharge of tympanic organs provides information concerning details of the behaviour of hunting bats. The discharge frequency gives an indication of distance; changes in frequency over a succession of bursts indicates whether the predator is closing in or moving away; and differences between the discharge from left and right tympanic organs provide information concerning the direction of approach. Precisely how these differences in input affect the pattern of motor output cannot at present be determined; but that they do, in fact, exert an appropriate influence is evidenced by the overt behaviour of moths in response to stimulation by artificial bat cries. Figure 10.3(b) shows the effect of stimulation by short. bursts of ultrasonic vibration on the flight path of moths. The source of stimulation, represented by a hatched circle, is mounted on a 14-ft mast, and two types of response are illustrated. One is the response to high intensities of sound (i and ii), which may be either a power dive, or a passive dive with wings closed. Here the direction of movement bears no relation to the source of stimulation, presumably because at high intensities the tympanic discharge saturates, and information about the

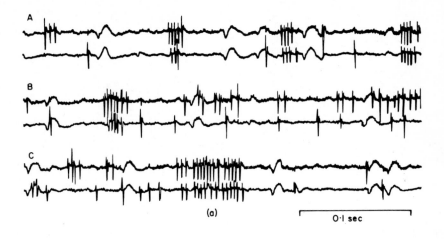

(a)

0·1 sec

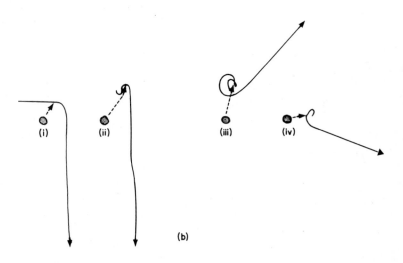

(b)

Fig. 10.3. Predator evasion of moths. (a) Impulse discharge in right and left tympanic nerves of a moth, in response to the cries of a red bat flying in the field. The slow waves represent depolarizations of the heart, and the large spikes which appear singly without synchrony are discharges from a third receptor associated with the tympanic organ, of unknown function. A, an approaching bat; differential response between right and left is marked at first but disappears as the bat flies overhead; B, a "buzz" registered mainly by one ear; c, a "buzz" registered a few seconds later by both ears (Roeder, 1963). (b) Flight paths of moths in response to an artificial ultrasonic sound pulse sequence. The hatched circle represents the source of stimulation. (i) and (ii), responses to high intensities of stimulation; (iii) and (iv), responses to low intensities of stimulation; dotted arrows mark the moment of stimulation. (Schematized from flash photographs of Roeder, 1963.)

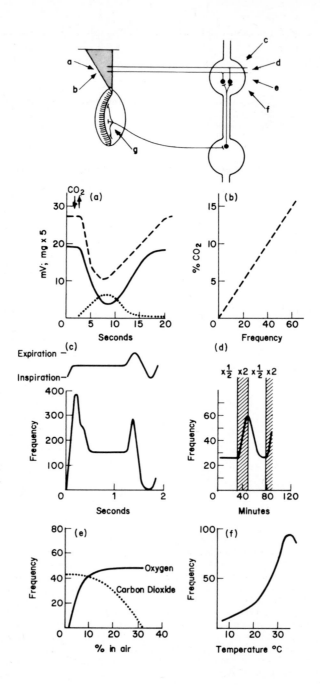

direction of the stimulus source would therefore be lost. At lower intensities of stimulation (iii and iv) it was possible to obtain clear indications of a directional component, with the moths taking a flight path directly away from the source of stimulation.

These experiments show that features of the sensory input are appropriately reflected in the pattern of motor output, even though it has not yet been possible to interpret the different output patterns, or to define the mechanism of integration, in terms of unit activity.

2. Patterns of Activity in Motoneurones

The one element of nervous architecture whose destination can be readily identified, and whose activity can be readily monitored, is the motoneurone, and a great deal of information has accumulated concerning patterns of activity in different types of motoneurone. Two examples have been selected for detailed discussion, to illustrate the sorts of results that have been obtained.

a. The Motor Supply to Spiracular Muscles

The neurophysiological basis of spiracular regulation has been under intensive investigation during recent years, and the results obtained furnish a convenient example of the effect of input, and of endogenous influences, on the pattern of activity in a motor unit. They have been schematized in Fig. 10.4, which will serve as a basis for the discussion that follows.

The motor supply to the closer muscle of a spiracle is shown as comprising two axons in the median nerve of the corresponding thoracic ganglion, the axons dividing immediately after they emerge from the ganglion to supply right and left spiracles. In the absence of ventilation these axons are continually active, firing at slightly different frequencies of about 10-20 impulses/s, so that recordings from the nerve show action potentials coming in and out of phase at regular intervals. This "free-running" pattern of activity gives rise to alternate periods of "fluttering" of the spiracular valve (see Chapter 7) coincident with

---

Fig. 10.4. Aspects of the neurophysiology of spiracular regulation. The site of action of various influences is indicated in the sketch, which shows the spiracle with its closer muscle on the left and two thoracic ganglia on the right. (a) The effect of a short burst of carbon dioxide, as marked by the arrows on: - - - - the magnitude of the muscle action potential; —— the tension developed during a twitch contraction; and . . . . the depolarization of the muscle membrane. (b) The relation between the frequency of motor impulses and the concentration of carbon dioxide required to open the spiracle. (c) The relation between impulse frequency and the different phases of the ventilation cycle. (d) Changes in impulse frequency recorded when the perfusion medium is switched from half strength to double strength of physiological saline. (e) The effect of oxygen and of carbon dioxide concentration on the discharge frequency. (f) The effect of temperature on discharge frequency. (g) Spiracular mechanoreceptor. Schematized diagrams from Hoyle, 1960(a) and from Miller 1964a b, 1965 (b-f).

synchronous firing, and of full closure when impulses arrive out of phase. This appears to constitute the background against which the control mechanisms of spiracular regulation operate.

The control mechanisms are of two different kinds, peripheral and central:

(1) the peripheral effects are based on a direct action of carbon dioxide on the closer muscle. The precise nature of this effect has not yet been fully elucidated, but carbon dioxide appears to cause a depolarization of the muscle membrane, which leads to a reduction in the magnitude of end-plate potentials, and a corresponding reduction in the force of contraction (Fig. 10.4(a)); and

(2) the central effects are mediated by changes in the frequency of discharge of the motoneurone, which affect the sensitivity of the closer muscle to the peripheral action of carbon dioxide. The higher the discharge frequency, the greater the concentration of carbon dioxide necessary to produce relaxation (Fig. 10.4(b)). A number of factors have been shown to influence the frequency of discharge of the motoneurone, of which the most important are:

(i) Ventilation. The motoneurones appear to be under the influence of excitatory and inhibitory interneurones, whose activity is linked to the cycle of ventilation, causing an increase in discharge frequency during expiration and an inhibition of discharge during inspiration (Fig. 10.4(c)). The details vary considerably from species to species, but the over-all effect is to synchronize the opening and closing of spiracles with the phases of ventilation, as described in Chapter 7. At the initiation of flight the motoneurone discharge is completely inhibited and spiracles open fully.

(ii) Water Balance. The state of water reserves has been shown to have a profound influence on the frequency of discharge in spiracular motoneurones, an effect which is thought to be mediated by changes in haemolymph concentration. The effect can be mimicked by perfusion of the metathoracic ganglion with saline solutions of different concentration, as illustrated in Fig. 10.4(d). Dilute solutions are associated with a low discharge frequency, while concentrated solutions produce high frequencies of discharge. On this basis, an insect whose haemolymph has been concentrated by dehydration would be expected to exercise stringent spiracular control, in agreement with the low rates of water loss which characterize insects in this state (see Chapter 7).

(iii) The Concentration of Respiratory Gases. If insects are exposed to gas mixtures containing subnormal concentrations of oxygen, there is a marked decrease in impulse frequency, which, in the well-hydrated insect, reaches zero, and thus ensures full opening of the spiracles, at a level of about 2% oxygen. Conversely, there is a decrease in discharge frequency as the carbon dioxide concentration increases above 10% (Fig. 10.4(e)).

(iv) Temperature. An increase in temperature produces a marked increase in discharge frequency over the range from $10°$ to $35°$, as illustrated in Fig. 10.4(f).

(v) Mechanical Stimulation. Mechanical stimuli applied to the region of the spiracular filter cause a discharge in associated mechanoreceptors (Fig. 10.4(g)). This produces an excitatory input to the motoneurone, and causes a burst of high frequency discharge to the closer muscle, and rapid closure of the spiracle. This presumably constitutes a mechanism by which the entry of dust particles or parasites is prevented.

Apart from the reflex response to mechanical stimulation, the precise nature of the effects described has not yet been established. Some of them are capable of being mediated by the isolated metathoracic ganglion, and could be ascribed to a direct effect of the factors concerned on the motoneurones themselves; on the other hand, it is possible that they may be based on afferent inputs originating in specific sense organs associated with the ganglion. The nature of ventilation control is also obscure; effects might be associated with the activity of stretch receptors monitoring the stresses developed during ventilation, or they could represent a collateral output from the motor centres controlling ventilation. Whatever the detailed interpretation, the results which have been obtained furnish a good example of the way in which the activity of motor units may be regulated by a variety of influences to produce a response that is homeostatic in the context of respiratory exchange and of water balance. Spiracular regulation appears to be governed predominantly by respiratory needs; it is carefully integrated with the phases of ventilation, and the peripheral carbon dioxide effect ensures that the needs of respiratory exchange are met according to the demands of the moment. The threshold of the carbon dioxide effect, however, is adjusted in the interests of water conservation, so that the frequency with which the spiracles open is reduced under conditions of desiccation. The temperature effect is thought to be related to the need for maintaining high thoracic temperatures for peak flight performance, but it would serve also to reduce losses of water that would otherwise occur at high rates from heated surfaces of the tracheal system immediately following flight.

b. The Coxal Adductor

A motor unit which shows interesting effects of a kind rather different from those described above is that which supplies the coxal adductor muscle of the jumping leg of locusts. This unit is concerned with the maintenance of posture, and it shows a regular tonic discharge in accord with this function. It has been shown that the frequency of the resting discharge is susceptible to changes of a relatively long-term kind, as the result of afferent inputs which are specifically related to its own output. If the frequency of motor discharge is monitored continuously, and if afferent axons of the tibial nerve are stimulated with a single shock every time the frequency of firing falls below a specific level, an increase in the mean discharge frequency is induced. By progressively raising the "demand" level, the mean discharge frequency can be more than doubled, as illustrated in Fig. 10.5, and the effect may persist for several hours after it has been produced.

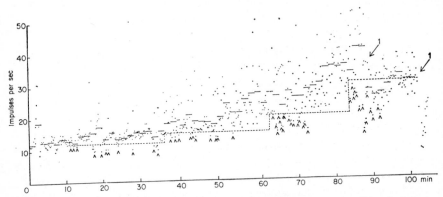

Fig. 10.5. Results of an experiment in which a shock was given to the tibia of a locust each time the mean discharge frequency over a 10-s period in the nerve of the right metathoracic anterior coxal adductor fell below a prescribed arbitrary "demand" level, indicated by the broken line. Moments of stimulation are indicated by arrows below the demand line. The dots show the mean frequency over a 10-s period, whilst the bars show the mean frequency over 100-s periods, to indicate the general trend more clearly. Note that after a few stimuli have been received the mean frequency rises. At very high demand levels an inhibitory effect appears to supervene (at arrows marked 1). (Hoyle, 1965.)

These results provide an adequate basis for interpretation of the original demonstration by Horridge (1962) of "learning" in a headless insect. This involved a preparation set up in such a way that the lowering of the metathoracic leg of a decapitated cockroach closed a stimulus circuit, and so delivered an electric shock to the lowered leg. It was found that after a relatively short period of time (10 min or so) the preparation "learnt" to avoid the electric shocks, by keeping the leg in a raised position. When this effect is divorced from the rather bizarre experimental situation used for its demonstration, and interpreted in the light of the experiments with locusts, it is clear that it may constitute a phenomenon of quite general importance. The results suggest that afferent inputs which follow systematically on significant trends in the motor output would work to produce a reversal of such trends. Mechanisms of this kind would constitute a satisfactory basis for the reflex adjustment of posture, for example, and as such would be of undoubted adaptive significance. Interest would centre on the precise neurophysiological basis of the underlying plastic change in nervous function, but it is unfortunately only too probable that the change occurs among the unseen activities of the smaller neuropile elements, and that it would therefore lie outside the range of experimental attack for the time being.

3. Central Inhibition

In an earlier part of this chapter the concept of motor centres was introduced, envisaged as neuronal networks capable of mediating an output, appropriately patterned in time and space, to a complex field. The anatomical basis of such hypothetical centres remains obscure, but the effects of operative procedures to be described in the present section add some support to the notion of their existence, whatever their precise nature. The free-running discharge of the single motor unit described above could be considered as representing one small fraction of the output from such segmental motor centres.

The most convincing demonstration of the activity of segmental motor centres comes from work on the reproductive behaviour of the praying mantis. One of the characteristic elements of the reproductive behaviour of the male of this species is an S-shaped bending of the abdomen, which serves to direct the terminal segments forward, and so permit the ovipositor of the female to be probed by the genitalia of the male. It has been shown that this motor pattern can be released by decapitation, or by transection of the ventral nerve cord at any point, and that it is the suboesophageal ganglion which, under normal circumstances, exercises an inhibitory influence over the lower segmental centres. It would seem that the sexual movements which occur in the decapitated mantis are the expression of appropriately patterned outputs from the abdominal ganglia, which might be produced endogenously, and without reference to sensory feedback. As mentioned in the introduction to this chapter, it is not possible to test this suggestion directly, by de-afferentation of the active centres, because of the mixed nature of insect nerves; but the concept of an endogenous origin for the patterned activity receives some support from the records illustrated in Fig. 10.6(a). These show the activity of motor units in a nerve passing from the last abdominal ganglion to the phallic apparatus, severed at a point distal to the electrodes, and with all other nerves to the ganglion cut to prevent input from segmental sense organs; the only connection between the ganglion and the rest of the nervous system is through the ventral nerve cord. The first part of the record shows the low level of discharge which occurs when the connection with the central nervous system, in particular with the suboesophageal ganglion, is intact. The second record shows an increase in activity 3 min after transection of the nerve cord, building up to the massive discharge of the last record, made after 7 min. If the records obtained over fairly long periods are examined, discharge frequencies in different motor units can be seen to wax and wane in a regular pattern, and it seems reasonable to suppose that these fluctuations represent the neurophysiological counterpart of the rhythmic copulatory movements which would have taken place had the nerves to the phallic musculature not been cut. If this interpretation is correct, it

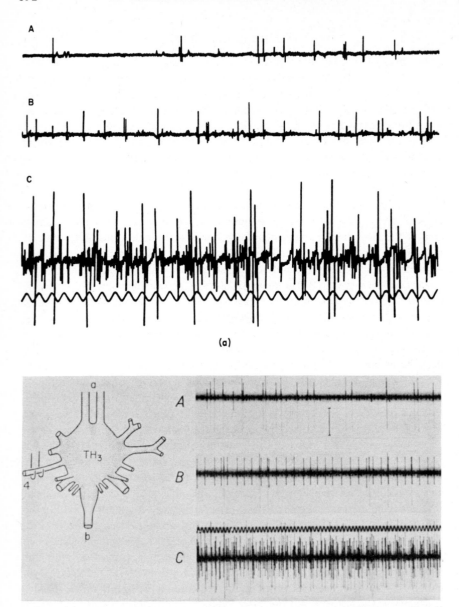

(a)

Fig. 10.6. The apparent release of motor centres from the inhibitory influences of higher centres. (a) Pattern of motor nerve impulses in the phallic nerve of the mantis. A, before connectives between the last abdominal ganglion and the rest of the nervous system have been severed; B, 3 min after transection of the nerve cord; C, 7 min after transection of the nerve cord. Time trace 100 cyc/s (Roeder, 1963). (b) Spontaneous efferent nerve activity in the isolated thoracic nerve cord of the cockroach. TH3, metathoracic ganglion with

demonstrates that the motor pattern is independent of sensory input, and lends support to the view of an endogenous origin.

In the praying mantis the release of sexual activity following decapitation is particularly strikingly manifested, and in view of the cannibalistic tendencies of the female mantis during courtship, which sometimes lead to decapitation of the male, it may well have considerable adaptive value. A similar phenomenon has, however, been demonstrated in the cockroach, where transection of the ventral nerve cord leads to the appearance of rhythmic bursts of motor impulses in nerves of the terminal ganglion. Here the corresponding motor activity is too attenuated in form to promote effective coupling, as it often does in the headless mantis, but there can be little doubt that the principle is the same, namely that of a segmental centre capable of a patterned endogenous output, but held under inhibition by higher centres. Nor is the phenomenon restricted to sexual patterns mediated by abdominal ganglia; Fig. 10.6(b) shows spontaneous activity in one of the nerves associated with the metathoracic ganglion of the cockroach, innervating coxal and other muscles of the metathoracic leg. In the presence of the head and the de-afferentated thoracic ganglia there is a low level of activity, but after connectives to the brain and suboesophageal ganglion have been cut there is a progressive increase in discharge rate, which reaches massive proportions after about 30 min. The change involves not only an increase in the firing frequency of previously active units, but also the bringing into operation of new units.

Inhibition of the activity of segmental motor centres is exemplified also in a grooming reflex of the locust, which has been investigated by Rowell (1965). The reflex can be elicited in a locust, restrained ventral surface uppermost, by a touch to the tactile receptors of the sternal region of the prothoracic segment. This elicits movements of the front legs directed towards the point of stimulation; both afferent and efferent pathways are confined to nerves of the prothoracic ganglion. In the intact preparation responsiveness is extremely low, and the reflex can rarely be evoked, but by a variety of operative procedures responsiveness can be greatly increased. If the nerve cord is severed anterior to the prothoracic ganglion, consistent results can be obtained, with the preparation responding about once in 10 trials. If, now, the input from lower parts of the central nervous system is progressively reduced, by de-afferentation of ganglia, or by severing the ventral nerve cord at different levels, there is a corresponding increase in responsiveness. The increase that results from isolating the prothorax from the abdominal chain is slight but significant; de-afferentation

---

connectives and lateral nerves; a, anterior; b, posterior; 4, lateral nerve innervating mainly coxal muscles and tergal and pleural depressors of the third leg. A, motor activity recorded from nerve 4 in a preparation that includes head and thoracic ganglia, without sensory input; B, the same 1 min after cutting the brain/suboesophageal connections; and C, 30 min later. Calibration 0.5 mV; time marker 50 cyc/s (Huber, 1965).

of the metathoracic ganglion now causes a marked increase in responsiveness, with positive reactions occurring approximately three times in every 10 trials. If the nerve cord is severed between meta- and meso-thoracic ganglia, the response rate rises to 8 in 10, while de-afferentation of the mesothoracic ganglion produces a preparation which responds all but invariably. Here appears to be another example of the release of a motor centre from inhibition, this time exerted by both higher and lower centres; and in this case the reflex appears to be steered in relation to the afferent input in such a way as to produce an appropriately orientated response, a condition which would be expected to apply to the operation of most motor centres under normal circumstances. One would imagine that the endogenous motor output, arising as the result of activity in a particular neuronal network, would produce a general pattern of movement whose finer details would be capable of adjustment on the basis of afferent input and feedback.

Experiments of the type described cannot provide more than an indication concerning the general principles of neuronal architecture. The operative procedures are too drastic to enable particular centres to be identified, since they usually involve the severing of an unknown number of fibre tracts of unknown origin and destination. A more promising approach to the elucidation of the details of nervous organization has been adopted by investigators using electrical stimulation of nerve centres in unrestrained animals, through chronically implanted electrodes. A motor pattern which has proved particularly amenable to this type of investigation is that involved in the sound production of crickets, based on the activity of stridulatory muscles innervated from the second thoracic ganglion. Three different types of song can be distinguished on the basis of differences in the pattern of sound—a calling song, a courtship song and an aggressive song. The performance of the corresponding motor patterns appears to be dependent on the activity of brain centres, and singing fails completely if connectives are severed between the brain and the thoracic ganglion. In this case there is clearly no question of a release of lower motor centres from the inhibitory influence of higher; on the contrary, there would appear to be an excitatory input from the higher levels.

The so-called mushroom bodies, or corpora pedunculata, of the supra-oesophageal ganglion seem to be closely involved with sound production in particular. They are situated one on each side of the protocerebrum, and consist of a massive neuropile receiving input from cephalic sense organs, and containing interneurones whose axons and collaterals appear to be restricted to the mushroom bodies. By electrical stimulation of certain parts of the neuropile, and of its associated fibre tracts, it is possible to evoke the calling and the courtship song (see Fig. 10.7); in other parts there may be an inhibition of singing during stimulation, while stimulation of the adjacent "central body" causes the production of atypical sounds. The performance of the different types of song is

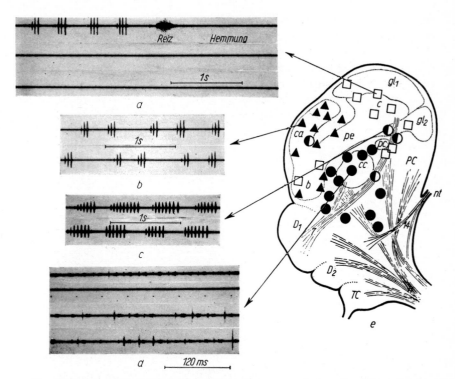

Fig. 10.7. Song patterns of male crickets and responses during stimulation of various parts of the brain. (a) Inhibition of calling by stimulation of the calyx of the mushroom bodies. (b) The calling song elicited by stimulation of the α-lobe. (c) The aggressive song elicited by stimulation in the tractus olfactorio globularis. (d) Atypical sounds elicited by stimulation of the central body neuropile. (e) Diagram of the cricket brain with the mushroom bodies and central body indicated by dotted lines and showing the distribution of points of stimulation. White squares, points of inhibition; black triangles, points at which calling is elicited; half-filled circles, points at which aggressive calling and associated behaviour is elicited; black circles, points at which atypical sounds are elicited; c, calyx; pe, peduncle; ca, α-lobe; pc, protocerebral bridge; cc, central body; PC, protocerebrum; TC, tritocerebrum (Huber, 1965).

accompanied by the postures and associated movements characteristic of the natural performance of the behaviour pattern, and the results fully confirm the long-accepted view that the mushroom bodies constitute important centres regulating the performance of complex patterns of behaviour.

This approach provides a means of identifying the nervous centres responsible for the regulation of motor activities much more accurately than can be done on the basis of ablation experiments. Unfortunately it represents no more than a first step toward an elucidation of details of higher integrative function; for substantial progress in this field a thorough knowledge of the distribution of dendritic fields and fibre tracts will be required, and technical advances must be made in relation to pin-point stimulation and recording. Here, as at the lower levels of nervous integration, the promise which is afforded by insects for the elucidation of the neuronal basis of behaviour, by economy of nervous elements, is counterbalanced by the technical difficulties posed by the nature of the insect neuropile, both for the neurophysiologist and for the anatomist.

CHAPTER 11

BEHAVIOUR

The study of insect behaviour has a history that stretches back over the centuries, and an enormous volume of literature has accumulated on the subject. Systematic investigation may be said to have started with the early naturalists, among whom the name of J. H. Fabre deserves special mention, but the investigation of insect behaviour was given a new direction and a fresh impetus during early decades of this century, under the influence of the simple mechanistic views which then prevailed. Insects, like other animals, were considered essentially as complicated machines, constructed to react to external stimuli in fairly simple and definable ways, and they were found to constitute particularly favourable material for experimental investigation of the relations between stimulus and response. Under the influence of this concept insect behaviourists proceeded to establish, in quantitative terms, the reactions of different species to a variety of physical factors, in a phase of investigation which may be said to have culminated with the publication in 1940 of "The Orientation of Animals" by G. Fraenkel and D. L. Gunn. By this time it was becoming clear that the behaviour of animals was not, in fact, capable of being convincingly interpreted in such simple terms, and that progress would have to depend on the formulation of a more satisfactory conceptual framework of interpretation. This was provided in the first place by results obtained from the study of behaviour in higher animals. These results suggested that what might be considered as the unit of behaviour was not so much a reflex as a movement, or "fixed motor pattern"; a movement which did not necessarily require an external stimulus for its evocation. They suggested, in other words, that what is of central importance in the investigation of behaviour is what movements an animal performs rather than the place to which such movements direct it; what it does rather than where it goes. The fixed motor patterns which, on this view, serve as the basis of behaviour may involve no more than the contraction of a single muscle, but more often they are constituted by a temporal and spatial pattern of activity in one or more motor fields giving rise, for instance, to the complex movements involved in walking or flying. The neurophysiological basis of these motor patterns has not yet been uncovered, but they may be considered

to arise from activity in neuronal networks whose basic structure is geno-
typically determined, corresponding to what in the previous chapter was referred
to as a motor centre. The corresponding movements may thus be considered as
inherited species characteristics in exactly the same sense as an anatomical
feature, such as the shape of a wing or the structure of mouthparts, is
characteristic of a species. This does not mean that the performance of the
movement is an invariable unit of behaviour, but simply that the animal
possesses the inbuilt potentiality for performing this type of movement;
precisely how the movement is made will depend not only on the structure of
the neuronal network, but also on the pattern of activity in it, and this in turn
will be a function of input from other parts of the nervous system.

If this "ethological" approach to behaviour, with which the names of
scientists like Lorenz and Tinbergen are particularly closely associated, can be
accepted as the most fruitful for the interpretation of insect behaviour, then the
first task that faces the student of insect behaviour is to analyse the complex
movements made by insects in terms of their constituent motor patterns, with
special reference to the functional significance of these patterns. Secondly it
would be necessary to consider the evocation of motor patterns in terms of
endogenous and exogenous influences; and to determine the extent to which,
and the mechanisms by which, the movements are steered in relation to features
of the environment, acting through the afferent system. Finally, consideration
would have to be given to possible modifications of the motor performance, or
of its afferent basis, associated with previous experience, in other words to the
existence of learning in insects. The account which follows has been split into
four main sections corresponding to these four aspects of behaviour: the motor
pattern, the evocation of motor patterns, the steering of motor patterns, and
learning.

1. The Fixed Motor Patterns of Insects

The morphological diversity of insects is rivalled, as one might expect it
would be, by the multiplicity of motor patterns manifested by different
members of the group, and detailed discussion would be beyond the scope of the
present volume. They range from simple movements like the "wing-waving" of

Fig. 11.1. Some elements of behaviour. (a) The postures of an ant worker cleaning itself.
A, cleaning right antenna and left foreleg; B, cleaning other legs; C, cleaning the posterior
part of the abdomen (Wallis, 1962). (b) Activity of the mealworm beetle at different
humidities; each triangle represents the proportion of animals active 15 min after induced
activation of all animals; the mean of each group is shown by the large circle (Gunn and
Pielou, 1940). (c) Examples of paths taken by the mealworm beetle in a gradient of
humidity with 97-100% relative humidity on the left and 94-97% relative humidity on the
right; black points show where the animals came to rest. The tracks give some indication
that random turning movements tend to keep the animals on the drier (right-hand) side.
(Gunn and Pielou, 1940.)

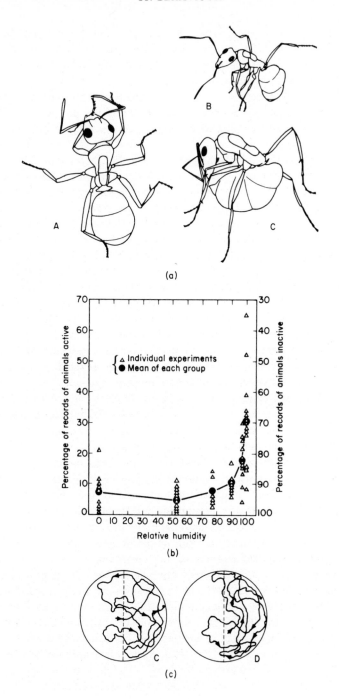

(a)

(b)

(c)

male fruitflies during courtship; through various grooming movements (see Fig. 11.1(a)), which may differ greatly in point of detail between species; to the co-ordinated sexual patterns of copulation, which have already been outlined in the case of the praying mantis; and from here to the more complicated movements involved in locomotion, whether walking or flying; and finally to the intricate patterns of motor activity operative during the performance of such special activities as the spinning of cocoons, the excavation of burrows or the construction of special nests.

It is clear from the examples given that specific motor pathways may be common to many different motor patterns; the movement of wings is involved both in courtship display and in flying; of legs in grooming and nest-building, as well as in walking. It is therefore necessary to conceive of the corresponding neuronal networks as showing extensive overlap, with elements of one pattern capable of being manifested as one of the components of another. Such an arrangement would ensure that different patterns could be placed at the service of a variety of different functions; walking and flying might in one instance be at the service of mating, at another at the service of nest-building or feeding; and each activity would be considered as operating within the context of broader categories of behaviour, e.g. reproductive, with the activity capable of being evoked from correspondingly diverse sources, as will be discussed below.

It should be emphasized that the interpretation here presented, of behaviour in terms of neurophysiology, of motor patterns as the expression of activity in complex neuronal networks, is quite hypothetical, and goes far beyond the available experimental evidence. It is adopted simply as an aid to a coherent presentation of the subject, and to emphasize the close relation that exists between the subject matter of this chapter and that of earlier ones, whatever the precise nature of that relation may eventually turn out to be.

One of the most characteristic features of the fixed motor pattern is the variation in intensity with which the movement may be performed, ranging from the barely recognizable "intention movement" to the full and vigorous performance. These differences presumably reflect differences in the level of excitation of the corresponding neuronal networks, based on variations in endogenous or exogenous input. In cases where locomotor patterns are involved, and where general physical features of the environment form an important source of input, we have an example of what, on the basis of earlier behavioural investigations, was defined as a "kinetic" mechanism. If, for instance, the linear velocity, or the frequency, of locomotion is affected by humidity (see Fig. 11.1(b)) the insect would be said to exhibit an orthokinetic response to humidity; if the angular velocity of locomotion is affected, the response would be of the klinokinetic type (Fig. 11.1(c)). Such responses have been demonstrated in a variety of insects, particularly in relation to temperature and humidity; they are of considerable biological importance, because they enable the insect to make an appropriate response to these important physical features, under conditions where an actual orientation to the stimulus might be impossible, because of the attenuation of its directional component. If, for instance, the gradient between humid and dry regions of the environment is so shallow as to preclude participation of a tactic steering mechanism (see below), the fact that the insect moves faster, or more frequently, in dry air than in humid air means that it will spend proportionately less time in dry regions, and aggregations will therefore occur in humid regions, which may be more favourable for survival. Where angular velocity is involved, similar aggregations may result, although the klinokinesis of insects requires a more searching experimental analysis than it has yet received for the unequivocal substantiation of its mechanism. It is significant, however, that kinetic mechanisms have developed in relation to the diffuse stimuli (temperature, humidity and smell), where the directional component is often too weakly developed for direct orientation.

2. The Evocation of Motor Patterns

Motor patterns differ greatly not only in their nature, but also in the type of situation with which they are associated. Some, like walking or flying, may be said to be of a general type, serving many different types of activity, and often appearing spontaneously, without obvious reference to changes in the stimulus situation. Others are highly specific, associated with particular functions such as reproduction, and only elicited in the presence of a particular stimulus, as for instance an object resembling a member of the opposite sex. It was in relation to these more specific types of motor pattern that the concept of the "innate releasing mechanism" was formulated, envisaged as the afferent analogue of the fixed motor pattern. The innate releasing mechanism can be considered as a sensory mechanism ensuring that the appearance of a particular configuration of sensory stimulation, the appearance of some definable proportion of the sum total of stimuli arising from the presence, for example, of a member of the opposite sex, leads to an input to the neuronal networks underlying a particular motor activity, say "wing-waving", and tending to its activation. The existence of this sort of coupling between innate releaser and fixed motor pattern, so that the presentation of a particular stimulus leads to the evocation of a particular activity, seems at first sight to accord well with a simple reflex interpretation, with behaviour seen as a sequence of motor activities induced by a corresponding sequence of specific stimulus situations. Elements of the reproductive behaviour of the digger wasp, for example, would be capable of interpretation in these terms. Care of the brood in this species involves a sequence of activities which include the digging of a nest; the hunting for caterpillars to provision the nest; oviposition; the bringing of more caterpillars and temporary closing of the nest, followed by a final closing when the nest has been fully provisioned. A sequence of complex motor activities is clearly involved in the total behaviour pattern, but certain fragments of the sequence could be interpreted in terms of reflex theory. The sight of the nest, for instance, would release the dropping of the caterpillar and scraping; the half-open nest would release digging; the open nest would release turning around, and the sight of the caterpillar would release the activity of pulling the caterpillar into the nest. A careful study of this behaviour pattern has made it clear, however, that motor activity is not evoked in any such simple way. To the digger wasp, the stimulus of a caterpillar may elicit quite different elements of behaviour depending on the context in which the stimulus is set, and on what may be termed the "motivation" of the wasp, which presumably reflects the distribution of excitation between different neuronal networks. If the wasp is hunting, the caterpillar is caught, stung and carried to the nest; but if a caterpillar lies close to the nest when the wasp is filling the entrance it may be used simply as filling material; or if a caterpillar is placed in the nest shaft when the wasp is digging out the nest, it will bring the caterpillar away, exactly as it would any other obstacle, such as a piece of wood. The same object thus releases quite different responses, depending on the state of the animal.

Locomotion and courtship movements, the general and the specific motor patterns, should not be considered as distinct in anything but a relative sense, representing perhaps the opposite ends of a continuous spectrum. The performance of specific patterns is usually triggered by the appropriate stimulus, but by no means invariably. In many cases they can be elicited by stimuli which bear no more than a faint resemblance to the normal, and sometimes they occur in the absence of any apparent change in the stimulus situation, as if spontaneously. Presumably the input from non-specific sources, endogenous and exogenous, has sufficed to raise the excitation of the motor centre to the point of discharge in the final common paths. Conversely, the general motor patterns may be elicited in response to specific stimuli, as walking may be evoked by the stimulus of a member of the opposite sex, as a component of courtship.

The performance of a given motor activity, the discharge from a given motor centre, and the intensity of that discharge, may then be taken as a reflection of the level of excitation in that centre, and this in turn will be a function of the level of input, inhibitory as well as excitatory, to that centre from exogenous and endogenous sources. A balance could be envisaged as existing between these two elements of input such that deficiencies in the exogenous component, inadequacies or absence of the releasing stimulus, may be compensated by high levels of endogenous excitation, while lack of endogenous motivation may necessitate the presentation of "super-normal" stimuli for elicitation of the particular activity.

a. Endogenous Input

The nature of endogenous effects can seldom be precisely defined in the present state of knowledge, and the possibility must be considered that several kinds of influence could be exerted on the excitatory state of a motor centre. There would certainly be effects associated with input from other centres, but in addition there might be direct effects of the chemical environment on the cells of the motor centre. Both could, in general terms, be described as reflecting the activation of the corresponding category of behaviour. For instance, onset of the reproductive phase of the life history would be associated with a variety of physiological and neurophysiological events which could potentially be effective in determining the level of excitation in a specific motor centre. The hormones regulating reproduction might exert a direct action on the membrane potential of particular types of neurone, and indications of such an effect have been obtained by experiments similar to the ones described in the previous chapter; the level of motor activity recorded from the nerves of the last abdominal ganglion of the mantis increases greatly during perfusion of the ganglion with extracts of the corpora cardiaca, in a way that mirrors the removal of inhibitory influences by cord transection. Similar direct effects may be involved in the changes of spinning behaviour which characterizes the wax moth at different stages of metamorphosis; it has been shown that implantation of corpora allata in last instar larvae induces the spinning of a larval cocoon instead of the pupal cocoon which would normally have been produced; and conversely, extirpation of corpora allata in early instars induces the formation of precocious pupal cocoons.

It should be mentioned that the regulation by hormones of different types of motor pattern is often rhythmical, so that the performance of the activity tends to occur at certain points in the 24-hr cycle. Where the motor pattern is locomotion, one gets the familiar diurnal rhythms of activity (see Fig. 11.2); where they involve movements associated with emergence of adults from pupae, one gets the eclosion rhythms, which have been demonstrated in a variety of species. Under normal circumstances the rhythmic endogenous influences appear to be set with reference to the rhythmicity of physical features of the environment, usually light or temperature, and diurnal changes in these tend to reinforce the endogenous rhythm; but the rhythmic activity persists for some time under constant physical conditions, although in the absence of reinforcement it tends to diminish in the magnitude of its manifestation, and to drift steadily out of phase with the 24-hr cycle.

The hormonal control of such diurnal activity rhythms has been convincingly demonstrated in the cockroach. When insects are maintained for a long time under constant conditions the rhythm eventually disappears; rhythmical activity can be re-established in such arhythmic individuals by joining them in parabiosis

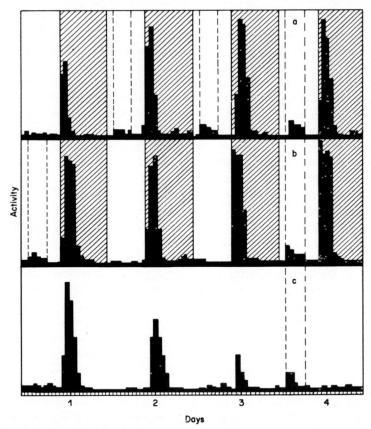

Fig. 11.2. Diurnal activity rhythms. (a) Activity of the cockroach in alternating light and darkness when food is obtained only in the light periods (shown between dotted lines). (b) Activity after three weeks of such conditions; during the second and third day food was not supplied, and the small activity peak normally associated with feeding is absent. (c) Activity after food has been introduced as usual for five days and the animals then starved in continuous light. Note that the rhythm continues for a few days, but becomes progressively less marked. (Harker, 1956.)

with another insect whose rhythm is still manifest, that is, by establishing continuity between the body fluids of the two insects. It appears that the hormonal secretions of the rhythmic individual, passing into the haemolymph of the arhythmic individual, are capable of inducing a rhythm in the second insect.

Changes in the pattern of metabolism associated with the growth of oöcytes in sexually mature females might constitute another type of endogenous influence. They could be associated with changes in haemolymph composition of a kind that might directly affect membrane polarization. In addition to such a hypothetical effect, and probably more importantly, the onset of reproductive activity would be associated with more specific nervous influences from other parts of the nervous system. The distinction between endogenous and exogenous factors would here tend to become blurred, because while some of the effects could be claimed as truly endogenous, those for instance which would result from changes in the level of discharge of stretch receptors associated with distension of the abdomen in pregnant females, others could be considered so only in the sense that they would be secondarily relayed from higher centres, rather than resulting from primary afference. The assumption of the reproductive phase would, in most insects, involve the activation of general behaviour patterns which in turn would occasion changes in afferent input. Early stages might be characterized by some form of appetitive behaviour, based on an activation of locomotor centres, leading possibly to quite specific changes in environment, hence of afferent input and of activity in corresponding neuronal networks, which in turn could exert an influence on the levels of excitation of more specific motor centres, destined for activation at a later stage in the sequence of reproductive behaviour.

At this point discussion becomes so general as to verge on the meaningless, nor would much be gained by an attempt to substantiate the occurrence of such effects with reference to specific examples. The case of a more specific and directly exerted endogenous effect should, however, be mentioned. This is the inhibitory input to a given centre which appears to be associated with the discharge of that centre. It is this phenomenon of negative feedback that forms the basis of the concept of the "consummatory act", the idea that the "goal" of a particular activity is not the attainment of some objective, but the performance of an act. The neurophysiological basis of this feedback inhibition has not yet been elucidated in insects, but it has been investigated extensively at the level of behaviour. It is usually found that when a given motor pattern has been elicited by the presentation of the appropriate stimulus situation, that same situation will fail to re-elicit the pattern if it is presented again, or will do so in a very attenuated form, despite the fact that care has been taken to prevent the achievement of the "goal" of the behaviour (e.g. feeding or copulation), which could markedly affect the general physiological state of the insect. The use of models have proved useful as a tool of investigation in this context. The sexual flight of grayling butterflies, for instance, can be elicited by female dummies. When such dummies are presented in succession, there is a decrease in the proportion of males that respond (see Fig. 11.3(a)) and a decrease in the intensity of their sexual flight. Similarly, the courtship behaviour of the parasitic wasp *Mormoniella* wanes rapidly if the males are presented with a succession of non-receptive females, which could be considered as analogous to artificial models. Not only do the models fail to elicit the courtship in a high proportion of trials, but when courtship does take place its intensity, as gauged by a variety of criteria, falls off sharply (Fig. 11.3(b)). There is no question that the

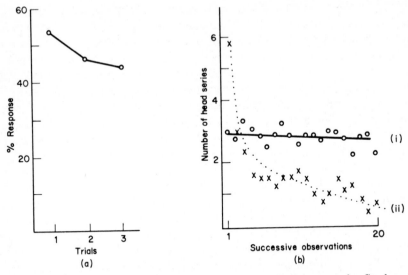

Fig. 11.3. The effect of repetitive stimulation on the performance of a fixed motor pattern. (a) The proportion of male grayling butterflies responding by sexual flight to successive presentation, at about 1½-min intervals, of a female dummy (Tinbergen *et al.*, 1942). (b) The number of "head series", a repetitive component of male courtship in the wasp *Mormoniella*, elicited in successive courtships with receptive females (curve i) and non-receptive females (curve ii), with 30-s intervals between presentation of the stimuli. (Redrawn from Barrass, 1961.)

extinction of the response can be attributed to neuromuscular fatigue, since other activities can be elicited at full intensity; it appears to be due to some kind of negative feedback associated with the performance of the behaviour pattern itself.

Under normal conditions the performance of a particular pattern of activity would be associated with the attainment of a "goal", with the fulfilment of a "purpose" as judged in anthropomorphic terms. The courtship of *Mormoniella* would lead to mating, the lowering of the proboscis of the blowfly in response to stimulation of tarsal chemoreceptors would lead to the intake of food, and so on. The corresponding changes in physiological state greatly affect the rate at which the behaviour patterns extinguish on repetitive stimulation. In *Mormoniella* copulation actually seems to have an excitatory effect on repetitive performance, so that when the males are presented with a succession of receptive females the complete behaviour pattern may be repetitively performed, and there is little waning in the intensity of the performance (see Fig. 11.3(b)). The feeding response of blowflies, on the other hand, elicited by contact of tarsal chemoreceptors with sucrose solution, extinguishes very slowly on repetitive presentation of the stimulus if the act of feeding is prevented. If, however, the fly is permitted to ingest a quantity of solution, the threshold for a positive response increases greatly, falling gradually during subsequent starvation to the original low values, as shown in Fig. 11.4. Here there appears to be little direct negative feedback from the performance of the motor activity, but changes in afference associated with the ingestion of food have an inhibitory effect on the performance of the motor pattern. This effect appears to be mediated by impulses from stretch receptors which monitor crop distension, passing via the

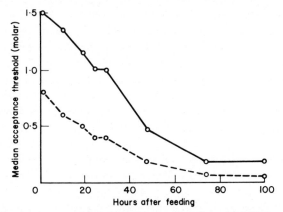

Fig. 11.4. Changes in sugar thresholds following ingestion of 2.0 M glucose; the full line gives the threshold for glucose, the dotted line the threshold for another stimulating sugar, fucose. (Evans and Dethier, 1957/8.)

stomatogastric system to the brain. Transection of the recurrent nerve prevents this indirect negative feedback, and the fly will continue to feed until it dies by distension.

The fixed motor patterns are thus susceptible to a variety of endogenous influences, both excitatory and inhibitory, the sum total of which determine whether or not the motor pattern will be performed, or the intensity with which it will be performed, under a particular set of external circumstances; and different types of motor pattern differ enormously in the degree of their susceptibility to such influences. Neither the neurophysiological basis of these effects, nor their site of action, have yet been elucidated.

b. Exogenous Input

(i) The Innate Releasing Mechanism. The nature of the effective releasing stimulus has been established for a great many motor patterns in a wide variety of insects, and a number of examples have already been given. They range from single component situations of great simplicity, as in the proboscis extension of the blowfly, where stimulation of a single labellar hair with a sucrose solution may be sufficient to initiate the complex motor response, and where only a single modality of stimulation may be involved. With other motor patterns a

number of different sensory modalities are involved simultaneously or successively, but the whole may still be at a relatively simple level, as in the reactions to prey of the water beetle, where water-borne vibrations cause release of the motor pattern, which in its final stages is dependent on visual stimulation of a rather unspecific kind. Even here the level of complexity is very considerable when one considers the enormous number of mechanoreceptive and visual sense organs that are involved in the total release. Releasing situations usually reach quite another order of complexity with motor patterns that mediate interspecific or intraspecific interaction, where the total stimulus situation may involve a number of different modalities of sensory stimulation—visual, olfactory, gustatory, mechanical—and where some of its components may in themselves be extremely complex. The sum total of stimuli that emanate, for example, from a female praying mantis, to impinge on mechanoreceptive, chemoreceptive and visual sense organs of the male during copulation would clearly be immense. It appears, however, that the releasing mechanism operates not on the basis of the total stimulus situation, but on what may be considered as an abstract of it, involving emphasis on certain of its elements while others appear to be ignored. For this reason it has proved possible to use appropriate models to release particular motor patterns, and by the use of such models to dissect the innate releasing mechanism into its constituent parts. An example of this approach has already been mentioned in connection with the sexual flight of grayling butterflies, which can be elicited with female dummies. By altering the properties of such models it has been possible to identify the major components of the releasing mechanism (see Fig. 11.5). The experiments indicate that shape is not an important element of the releasing mechanism, since models whose outline give a fairly realistic impression of a butterfly are no more effective in eliciting a response than circles or even rectangles, provided the disproportion between length and breadth is not too great; nor does colour seem to be of much importance in the context of sexual pursuit, although it constitutes a very important element of the releasing situation for another behaviour pattern, that of feeding. The factors which were found to be of special relevance to the release of sexual pursuit were reflectivity, size and nearness, and the type of movement with which the presentation of the model was associated, whether it was moved smoothly across the visual field of the male, or in an irregular way. The optimal model was a relatively large object, as dark as possible, as near as possible, and moved in the typical fluttering way of a butterfly's flight.

Experiments of this kind provide a convincing indication that certain elements of the total stimulus situation have been endowed with special significance for the release of a given motor pattern. Such discrimination is presumably based on special configurations of afferent and interneuronal pathways, such that the pattern of activity induced by these elements is especially effective in raising the level of excitation in the appropriate motor

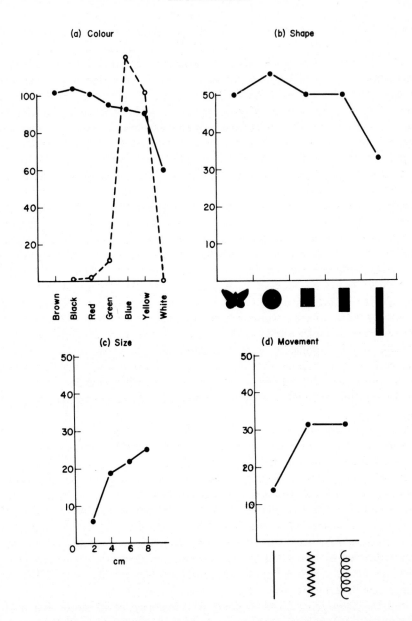

Fig. 11.5. The composite nature of the releasing stimulus, analysed by means of models; the graphs show the proportion of male grayling butterflies responding by pursuit to the presentation of female dummies. a (solid curve), of different colour; b, of different shapes; c, of different sizes; d, moved in different ways. The dotted curve in (a) shows the effect of colour on the feeding response (redrawn from Tinbergen *et al.,* 1942).

centre. In view of the artificial nature of the experimental situation, and of the variability involved in behavioural experiments, one could not conclude from such experiments that any element of the normal stimulus situation is totally ineffective in contributing to motor excitation. It is sometimes possible, however, to carry out experiments in the context of a relatively unaltered and natural situation, and in this way to get more reliable evidence of the selective basis of the release. One of the most striking examples of this is the releasing stimulus for the act of copulation in the parasitic wasp *Mormoniella,* which is of classical simplicity. It consists of a lowering by the female of its horizontally extended antennae to a depressed position, out of reach of the male's mouthparts. Various other movements of the female are involved, but this antennal depression appears to be an indispensable condition. It can be simulated by antennectomy, and under these circumstances the male regularly attempts to copulate, but without success because the female fails to perform the movements normally associated with antennal depression, and essential for successful coupling. This example is of particular interest in that the releasing stimulus involves an absence of a stimulus, the female antenna, that was formerly there, indicating that certain stimuli may be inhibitory to the release of a motor pattern, in other words that the releasing mechanism is not based solely on excitation. Similar situations have been described for a number of other species, as for example in certain grasshoppers, where courting behaviour is inhibited by the songs of rival males, or in butterflies, where male markings, black with a white dot, inhibit the courting reactions of other males.

The fact that the ability of certain stimuli to evoke certain motor patterns is an inherited feature, that the innate releasing mechanism is really innate, has been convincingly demonstrated in a number of cases. Insects which have been raised in isolation are capable of responding to the complex succession of stimuli involved in courtship and mating; bees which have been raised in an incubator, after witnessing the tail-wagging dance of another member of the species indicative of the locality of a food source (see below), are able to fly, first time, in the indicated direction and for the indicated distance. This is not to say that no learning is involved in the discrimination of particular stimulus situations, but it does show that the basis of such discrimination is innately given as the counterpart of the fixed motor pattern, and, like that, presumably represents some particular configuration of neuronal architecture, but on the afferent rather than the efferent side.

(ii) The Steering of Motor Patterns. The performance of motor patterns is often orientated in relation to some feature of the total stimulus situation. Locomotion, for instance, elicited by the stimulus of a member of the same species as part of aggressive or reproductive patterns of behaviour, will be directed towards the releasing stimulus. Or locomotion, evoked endogenously, may be orientated in relation to a source of light, away from it or towards it or

at some specific angle to it; or in relation to gravity or to a gradient of humidity. The adoption of some particular posture may be similarly directed in relation to some particular source of stimulation, as when locusts bask in the rays of the early morning sun; or in relation to some particular object, as in the posture adopted by the praying mantis in relation to its prey immediately before the strike. Whatever the precise relation between the orientating stimulus and the motor pattern, and whether the orientating stimulus constitutes a part of the releasing situation, or is indifferent in relation to the evocation of the motor pattern, it is clear that there are basic mechanisms which enable the performance of motor patterns to be steered in relation to environmental stimuli, and the question arises as to the precise nature of these mechanisms.

It is in this context that the immense amount of work carried out during the early years of behaviour investigations, under the influence of the reflex approach to the study of behaviour, finds its rightful place. The results were usually based on a study of the reactions of insects to simple physical stimuli such as light, temperature, gravity, humidity, etc., but there is no reason to suppose that the visual orientation to a point source of light, for instance, differs in essence from orientation with reference to a member of the opposite sex, or an object of prey. What is important is not what is the object of orientation, but what is the mechanism by which orientation is achieved in relation to that object, and this has been established by such early work for a number of different types of reaction, in behavioural if not in neurophysiological terms.

The nature of the orientation mechanism appears to reflect in some measure the properties of the orientational stimulus. It has already been noted that with diffuse stimuli, for which a directional component may be lacking or weakly developed, kinetic mechanisms, characterized by an absence of orientation, may be of importance. When directional components, in the form of gradients of stimulation intensity, are sufficiently strong, orientation to such stimuli becomes possible, and it is usually achieved on the basis of a comparison of stimulation intensities, by a mechanism referred to as tropotactic. In many cases the comparison on which orientation is based involves bilaterally symmetrical receptors, so that an element of balance appears to be involved; if one receptor field is more strongly stimulated than the other, appropriate turning movements are initiated, and they continue until equality of stimulation is achieved, at which point the animal will be orientated with its long axis parallel to the gradient, and will move away from, or towards, the source of stimulation. The response of bees to attractive odours furnishes a useful example of tropotactic behaviour (Fig. 11.6(a)). Tested in a Y-tube olfactometer, normal bees will turn towards the source of attractive odour at the Y-junction; if, however, the antennae are immobilized in such a way that their tips are separated by less than 2 mm, they are unable to differentiate the simultaneous comparison between right and left, on which orientation is based, and they move

se

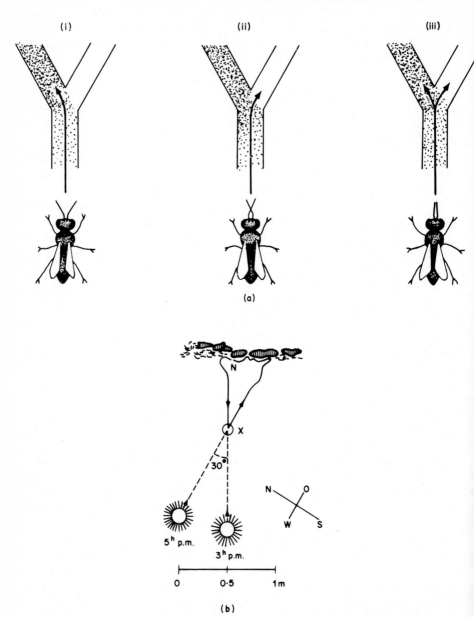

Fig. 11.6. Examples of the steering of motor patterns by environmental stimuli. (a) The reactions of bees to attractive odours. (i), normal bees tested in a Y-tube olfactometer select the air stream that carries the odour; (ii) bees, whose antennae have been fixed in such a way that the right antenna projects on to the left side, the left antenna on to the right side, select the odourless air stream; (iii) if the antennae are fixed in such a way that antennal tips are separated by less than 2 mm, the bees move indifferently into air streams with and without odour. (Data from Markl and Lindauer (1965.) (b) The "sun-compass" orientation of ants. At the point marked x, the ant was confined in a dark box for 2 hr. When freed, it assumed, in accord with the changed position of the sun, an incorrect direction on its way

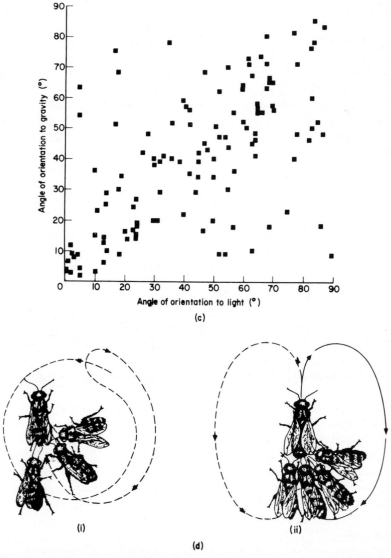

(c)

(d)

back to the nest, N (Markl and Lindauer, 1965 from von Frisch). (c) The graph shows the relation between the angle at which an ant was orientated to light, and the angle at which it subsequently orientated to gravity, when the light was switched off, and the platform swung from the horizontal to the vertical position; the angles were measured as the smallest angle between the track and the orientating stimulus, irrespective of direction. The correlation coefficient is 0.587, significant at the 0.001 level of probability. (Vowles, 1954.) (d) "Dances" of the honey-bee. (i) The pattern of the round dance. The dancer reports a source of food close to the hive. The dance followers maintain contact with the dancer by means of the antennae. (ii) The pattern of the tail-wagging run. The straight run between the two semi-circular ones is emphasized by wagging the abdomen. The straight tail-wagging run contains information about the location of the food source. (Lindauer, 1965, after von Frisch.)

indifferently into either side arm. If the antennae are fixed in such a way that the left antenna extends over to the right side, and the right to the left, the insects turn regularly into the odourless branch. It would seem that at the Y-junction the left antenna, in its artificial position, is less strongly stimulated than the right, and initiates turning movements which take the insect away from the source of stimulation.

Tropotactic mechanisms, involving an element of balance between symmetrically disposed receptor fields, have been demonstrated for humidity and temperature reactions, as well as for reactions to smell. Failure of orientation following unilateral extirpation of a sensory field, often coupled with so-called circus movements, which involve a continual turning towards, or away from, the intact side, will often provide evidence of the existence of tropotactic mechanisms. On this basis the orientation to light of certain insects appears also to involve a tropotaxis, but more usually reactions to stimuli with a strong directional component do not have to rely on comparisons of this type.

The neurophysiological basis of tropotactic orientation has not been elucidated, but it is one which can be envisaged in fairly simple terms. One could postulate that differences in the intensity of stimulation at bilaterally symmetrical receptors would be associated with corresponding differences in the level of input to bilateral centres of locomotion; this could, in turn, find expression in differences in the amplitude or frequency of limb movement, differences in "muscular tone" as postulated by earlier workers, to produce the appropriate turning component.

It is possible that orientation based on comparisons that are successive in time, rather than simultaneous, may be involved in the orientation of many insects to diffuse stimuli, although unequivocal experimental evidence is lacking. It is difficult to imagine that the known orientation of insects in flight to a source of smell could be achieved on the basis of a simultaneous comparison of stimulation intensities at bilaterally symmetrical receptors, in view of the likely effect of turbulence on the minute gradient which could be presumed to exist between right and left antennae. Under these circumstances a comparison of stimulation intensities at different points of the gradient, as it is traversed by the insect during flight, would appear to furnish a more promising basis for orientation.

Reaction to stimuli with a strongly developed directional component, such as light or gravity, are usually mediated by mechanisms referred to as telotactic, where the insect "fixates" the source of stimulation and moves away from it or towards it, or at some specific angle to it; in the latter case the reaction is referred to as a compass reaction or menotaxis. Compass reactions were discovered initially by observations of the behaviour of ants, which were found to maintain a fixed course during their foraging activities, by moving at a certain angle to the sun's rays. If the ant was confined in a black box at some point along its route, and released after an interval of time, it would take up a course at an angle to its original course equal to the angle through which the sun had moved during the intervening period. Or if it was trapped at the end of its foraging run, it would set off for the return trip to the nest in a direction that showed a corresponding deviation from the correct one (see Fig. 11.6(b)).

Similar orientations in relation to gravity have been demonstrated in a number of insects. Ants, for example, can be trained to move to food from the centre of a vertical turntable, and to return to the centre, along any angle to the vertical, in the absence of light; only slight deviations from the target occur. Here the body is kept at a constant angle in relation to gravity rather than in relation to light.

Reactions of the type described obviously involve very much more than a simple reflex response to stimulation, explicable in simple neurophysiological terms. In the case of the light compass reaction, for instance, continual adjustments have to be made to the angle of orientation to compensate for movements of the sun, adjustments which fail to be made when the insect is imprisoned in the dark. Again, for both the visual and the gravity orientated responses, orientation can be switched through $180°$ for the return journey. The complexities involved are further emphasized by the remarkable ability of many insects to transpose a compass course from light to gravity and vice versa. The angle that is maintained in relation to a light source during locomotion on a horizontal surface is kept constant in relation to the gravitational force in the absence of light, when the surface is changed to a vertical one. For instance, the "escape run" of ants may be orientated in relation to light on a horizontal surface or to gravity on a vertical surface. If, during a run on a horizontal surface, the light is switched off and the surface swung to the vertical position, the run will continue at an angle to the gravitational field equal to that which previously obtained in relation to light (see Fig. 11.6(c)). A more natural example of such transposition is provided by the behaviour of foraging bees, following discovery of a source of nectar or pollen. On their return to the hive they communicate their discovery to other members by performing a special dance, the pattern of which may provide information about the direction in which the source is to be found, and about its distance from the hive. If it is within a hundred metres of the hive a "round-dance" is performed (Fig. 11.6(d)(i)), which signifies simply the presence of a source in close proximity. If the source is further away a "tail-wagging" dance is performed (Fig. 11.6(d)(ii)), in which the frequency with which the tail-wagging run is performed provides a measure of distance, while the orientation of the run on a vertical surface in relation to the gravitational field gives an indication of the direction of the source in relation to the sun.

The complexity of the mechanisms by which motor patterns are steered in these examples precludes a convincing neurophysiological interpretation in the present state of knowledge, nor can there be much hope of bridging the gap between neurophysiology on the one hand, and behaviour on the other, until technical advances make possible an exploration of electrical activity in the neuropile in relation to functional organization. In the meantime, empirical studies of behaviour can do much to provide the firm foundation of quantitative information which must serve as the ultimate basis of a unitary interpretation. An outstanding example of the progress that can be made is furnished by the work of Mittelstaedt on the prey-catching strike of the mantis, as interpreted on the basis of information theory (Fig.11.7). When an insect moves within striking distance of the mantis it is visually fixated, the head of the mantis moving until

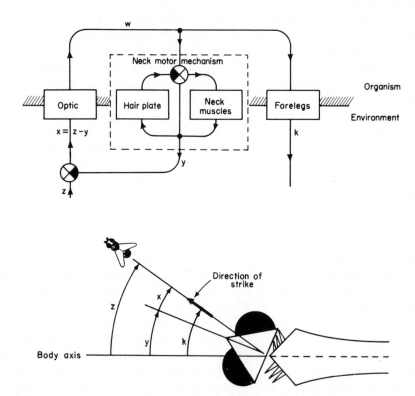

Fig. 11.7. Control pattern of prey localization in mantids. The prey is in a position which deviates from the body axis by the angle z, as monitored by the optic subsystem; the output from this system is an input variable, w, or "order", to the neck motor mechanism; this causes a movement of the head towards the prey, continuing until the deviation y, of the head from the body axis is equal to the deviation of the prey from the body axis; the deviation $x = z - y$, of the prey from the eye axis is then zero, and the insect is fixated. At this point, the "order" to the neck motor mechanism will be accurately proportional to the deviation of head, and hence of prey, from the body axis, and this "order" is used to determine the deviation, k, of the strike from the body axis. (Mittelstaedt, 1962.)

the image of the prey stimulates both eyes equally. At this point the strike is made by the forelegs, which are extended to the right distance and in the right direction to grasp the prey. The fixation is a relatively slow process, and appears to be controlled with little inertia or oscillation by a circuit of two closed loops each of two components; one between the visual mechanism and the motor mechanism of the neck; the other between hair-plate proprioceptors on the neck and the neck muscles. The strike, on the other hand, which is performed with lightning rapidity, and may be completed in less than 30 ms, is controlled by a single motor mechanism without feedback. There is, in other words, no mechanism for correcting errors in the strike once it has been released; its direction and extent is determined by steering mechanisms operative during the adoption of the strike posture, and these mechanisms in turn are based on input from visual and proprioceptive receptors. It is obvious that such detailed analysis of the components involved in the steering mechanism, and of their quantitative interrelationship, will serve as an indispensable basis for an eventual analysis in neurophysiological terms.

3. Learning

The extent to which the behaviour of insects is influenced by previous experience, the extent to which insects learn, has been much investigated. With insects, as with other animals, it is possible to distinguish a number of different kinds of learning, of which the simplest is the phenomenon of habituation to which reference has already been made. It is exemplified by the escape behaviour of the cockroach which, in the face of repetitive stimulation "learns" to ignore the stimulus; and in broader terms it finds expression in the general tendency for motor patterns to extinguish under conditions of repetitive stimulation. The neurophysiological basis of these effects have not been unequivocally established, and a number of different causes, including sensory adaptation and negative feedback from activated motor centres, may contribute to the total effect. This sort of learning is often of fairly brief duration, the memory short, though this is by no means invariably the case.

Other types of learning which are well exemplified by the behaviour of insects are "associative learning", which involves the building up of conditioned responses, and "trial-and-error learning". Many insects have been shown to be amenable to training of one kind or another, involving usually the association between stimuli which in themselves do not release any kind of response, with situations that do. Bees, for instance, can be trained to extend their prosbosces in response to odour, and are able to associate the colour of feeding places with the presence of food. Complex configurational stimuli can also be used as cues for food, and many insects have been shown to possess a high capacity for route learning in artificial mazes, a variety of stimulus modalities being put to use in identifying the correct route.

The olfactory conditioning which has been demonstrated in several species of insects bears a curious resemblance to the process of "imprinting", which is a feature of the learning process in many vertebrates. Here the full releasing stimulus is not innately given, but is in large part determined by experience during a particularly "sensitive" period of the life cycle. In the same way, the food or host choice of certain insects appears to be determined during early developmental periods by the substrate on which development occurs; transfer of choice from the normal host plant to an unusual one can be effected by raising the larval stages on the unusual type.

Evidence of a capacity for so-called "latent learning" is available from observations of the behaviour of insects in their natural environment, particularly in the context of route finding. The ability of the digger wasp to provision its nest with a succession of stung caterpillars, for example, is based on a capacity to learn the location of the nest, so that it can be found at the end of successive foraging trips. Such learning appears to be based on responses to a complex combination of features of the environment, and it seems that many insects build up a detailed spatial knowledge during preliminary orientation flights.

The one form of learning which has not been demonstrated in insects is so-called "insight learning", defined as the adaptive reorganization of the present content of experience in what may be described as "intelligent" behaviour. In view of the prevalence of other forms of learning in insects and of the fact that, with the exception of simple forms of habituation, the neural substratum of memory and learning in insects and in other animals remains obscure, it would perhaps be premature to dismiss the possibility of insight learning in insects. Even as they are, the demonstrated capacities for learning in the group are surprisingly extensive, and accord ill with the widely accepted concept of insects as the prime exemplars of a form of life based on innately determined behaviour. It would seem that despite the short life span which characterizes most members of the group, and despite the relatively small number of neural elements incorporated in their central nervous system, experience may play a prominent part in moulding behaviour patterns whose basis is innately given.

4. Conclusions

The question arises whether it is possible, against the background of this brief review of the main aspects of insect behaviour, to recognize features which could be interpreted as characteristic of insects as a group, and which could be seen as counterparts of other peculiarities of insect organization. The widespread occurrence of reactions to humidity could perhaps be seen as such. In terrestrial animals as small as most insects are, the threat to water balance posed by the terrestrial atmosphere is a serious one, and humidity will clearly constitute one of the most important physical features of the environment. The subject will be discussed further in the last section of this book, and all that needs to be said here is that the possession of well-defined reactions to humidity, mediated by a variety of receptor types, effected by a variety of reaction mechanisms, undirected as well as directed, and regulated in a variety of ways, may be considered to represent a general characteristic of the group.

In considering other features of insect behaviour which could be classed as general characteristics, one may revert to the widely accepted view of the essentially instinctive nature of insect behaviour, and set against this the mounting evidence of a learning capacity which is little if at all inferior to that of other animal groups, not excluding vertebrates. If a substantial component of the behaviour of insects is in fact based on previous experience, then one may ask how the concept of insects as essentially instinctive has arisen. It seems possible that its source may lie in the apparent automatism of much of the motor activity of insects, which often has a robot-like appearance. This, however, may have little to do with whether or not the motor pattern is innately determined; what it may reflect rather is a characteristic which could stem from quite another feature of nervous organization, namely the lack of a strongly developed element of feedback. Insects are small animals, and their motor patterns are performed correspondingly quickly and at correspondingly high frequencies, as witnessed by the 30-ms prey-catching strike of the mantis, or the 0.5-ms wingstroke of the fly. Coupled with this, the nerve fibres of insects are unmyelinated, and therefore characterized by relatively low conduction velocities. It seems likely, therefore, that phasic feedback would in many cases be ineffective in producing corrections to a motor act, since the movement would have been completed before the appropriate information could be relayed through the central nervous system. As with the strike of the mantis, the movements would have to be largely predetermined, because by the time information is available concerning possible error, it will be too late to correct them. It may be this lack of fine adjustment of motor performance in relation to features of the environment, firmly imposed by properties of the nervous system, that produces an appearance of automatism; but such automatism does not imply a prevailing innateness, or an absence of learning; it is a simple defect of the motor act, imposed by the discrepancy between the speed with which the act is performed and the time required for effective feedback.

This concept receives some support from recent investigations of the control of flight in locusts, which have shown that the input from proprioceptors in the wings are not effective as phasic feedback, but that the information is integrated over a number of wing-beat cycles to produce a gradual adjustment in wing-beat frequency (Wilson, 1965). Another example of apparent lack of phasic feedback comes from observations on the grooming behaviour of certain insects. The wing-cleaning movements of flies normally consist of rubbing the hind leg over the surface of the wing; if flies have their wings amputated at emergence, the normal leg movement occurs, indicating that the motor pattern is "fired off" and proceeds to completion in the absence of appropriate feedback. The suggestion is not intended that lack of feedback adjustment is a universal feature of insect motor patterns, and indeed, phasic feedback has been shown to play an important part in the performance of many motor patterns; but in some cases it is precluded by the rapidity with which the movement is made, and it would appear to play a relatively minor part in the motor patterns of insects as compared with higher animals. It seems possible that this may represent a general characteristic of insect behaviour, reflecting an aspect of the organization of its nervous system.

The Physiology of Reproduction and Development

INTRODUCTION

Two of the general characteristics of insect organization are of particular relevance to the physiology of reproduction and growth. One is the terrestrial mode of life, which has necessitated the development of internal fertilization and the provision of a strong and well water-proofed eggshell, and some of the special features of the reproductive system can be seen to provide a reflection of these requirements. The second is the possession of a rigid exoskeleton, which is of special significance in relation to the problem of growth; with such a skeleton, the animal is effectively contained within a cuticular prison of relatively fixed dimensions, and an increase in size can only take place by replacing one cuticular box with a larger one. The growth of insects tends, therefore, to be a discontinuous process, achieved in a series of steps marked by the shedding of the old cuticle in the process known as ecdysis. This stepwise transformation of small individuals into bigger individuals appears to have provided a suitable physiological basis for the far more spectacular transformations of metamorphosis, which characterizes many of the advanced orders of insects. Here the change is not from a smaller to a bigger individual, but from an individual of one kind to an individual of quite another kind, as in the transformation of a fly maggot into its winged adult, through the intermediary of the pupal stage. Here the individual organism is strikingly polymorphic, passing in the course of development through a succession of different forms, and the study of the mechanisms by which this metamorphosis is achieved has proved to be one of the most exciting in insect physiology.

CHAPTER 12

REPRODUCTION

The main features of the reproductive organs of insects are illustrated in Fig. 12.1. The testes and ovaries are of mesodermal origin, while proximal parts of the reproductive system in both sexes derive from ectodermal invaginations, and

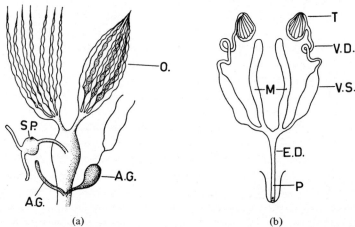

(a)   (b)

Fig. 12.1. The reproductive organs of insects. (a) Diagram of the female reproductive system (de Wilde, 1964 after Weber). (b) Diagram of the male reproductive system (de Wilde, 1964 after Wigglesworth). A.G., accessory gland; E.D., ejaculatory duct; M, mesodenia, O., ovarium; P, penis; S.P., spermatheca; T, testis; V.D., vas deferens; V.S., vesicula seminalis.

are provided with a cuticular lining. In most species the tracts of both sexes carry prominent accessory glands; in the female they secrete a variety of adhesive substances which serve as protective coverings for the egg, and in the male they produce, often in co-operation with glandular parts of the genital tract, the seminal fluid and the special sperm capsules, or spermatophores, in which the sperm is transferred to the female. In addition to the accessory glands, the female has one or more diverticula, known as spermathecae, in which sperm are stored after insemination.

a. Oögenesis

The ovary consists of a number (up to 2000) of egg tubes or ovarioles; the primordial germ cells are situated at the distal ends of the tubules, and as the oöcytes mature and grow, they move along them towards the oviduct. Three main types of ovary can be distinguished on the basis of oöcyte nutrition (see Fig. 12.2). In the panoistic type there are no special nutritive cells, other than the somatic follicular cells which surround each egg. In the polytrophic ovary, some of the undifferentiated oögonia develop to form nurse cells, or trophocytes, which contribute to the nutrition of the oöcyte during early stages of development, and accompany it on its passage down the ovariole. In the telotrophic type, the nurse cells do not accompany the oöcyte in its descent towards the oviduct, but remain at the apex of the ovariole; the nutritive materials which they supply are conveyed to the oöcyte through special nutritive cords.

During the process of oögenesis, materials required for subsequent embryonic development are incorporated in the substance of the egg cell as "yolk". The materials mainly involved are proteins, fats and carbohydrates, and they appear to be supplied by the trophocytes and follicular cells, taking the appearance of yolk spheres after they enter the oöcyte. Substantial quantities of nucleic acid must also be provided for the growth of the embryo, and in panoistic ovaries this appears to be supplied by extrusion of granules known as "chromidia" from the nucleus of the oöcyte itself. In the other types, the oöcyte may play little or no part in the synthesis of nucleic acids, which derive instead from the nurse cells.

The sequence of events involved in the vitellogenesis of a polytrophic ovary is illustrated in Fig. 12.2(b), which shows the early differentiation of trophocytes from oöcyte; the accumulation of chromidia coinciding with peak activity of the nurse cells; the regression of nurse cells during peak activity of the follicular cells; and the simultaneous appearance of fat droplets in the egg cytoplasm

The last stages in the formation of the mature egg involve the laying down of egg membranes. The thin vitelline membrane forms at the surface of the yolk, and on this is deposited the eggshell, or chorion, as a cuticular secretion of the follicular cells. The covering laid down by this single layer of cells rivals in complexity the cuticle itself, being composed in some insects of as many as seven distinct layers (Fig. 12.3(a)); the layers are distinguishable on the basis of composition, and proteins, lipids, polyphenols and mucopolysaccharides are principal constituents. It provides substantial mechanical protection, but its deposition at this stage raises a problem in relation to fertilization, because a massive shell of this sort would clearly provide an impassable barrier to the passage of spermatozoa; and it would also militate against an adequate supply of oxygen during the later stages of development when respiratory rates are relatively high. Both of these problems have been solved by the provision of a micropyle, a region of the egg surface at which the chorion is either sufficiently

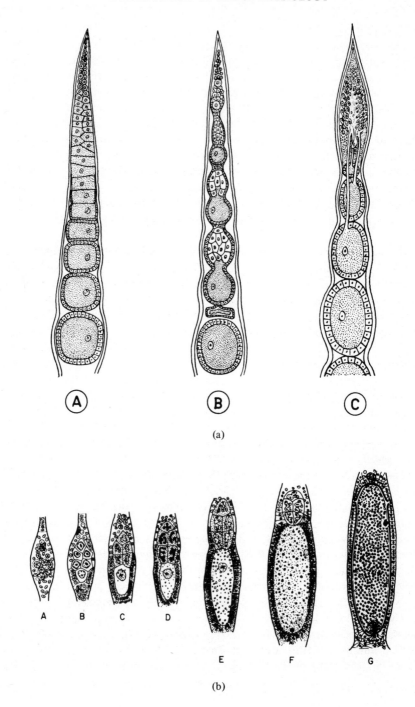

(a)

(b)

thin to allow ready access of oxygen and of sperm, or where there are actual perforations of the eggshell.

The chorion of many species of insect has recently been shown to be intricately sculptured at the submicroscopic level as well as at the microscopic level, a phenomenon which appears to represent a respiratory adaptation of a rather paradoxical kind. One would normally think of dryness as a characteristic of the terrestrial environment, forgetting that such environments are subject to periodic wetting by rain or dew, and that what might be required for the eggs of many terrestrial insects, fastened to the substratum, and therefore liable to prolonged inundation, would be adaptations for aquatic as well as for terrestrial respiration. The sculpturing of the eggshell does in fact appear to represent a special device for ensuring adequate respiratory exchange during periods of immersion. It usually takes the form of one or more networks of chorionic processes, whose meshes entrap continuous and intercommunicating layers of air (Fig. 12.3(b)). The dimensions and physical properties of the system are such that it presents a strongly hydrofuge surface, and when submerged in water it will therefore constitute a physical gill, or plastron, providing a large air/water interface across which oxygen can diffuse to supply respiratory demands.

There remains to be considered the water-proofing of the insect egg; lipid components have been shown to be incorporated in the chorion, but they do not appear to constitute an effective barrier to the diffusion of water, for at the time when the development of the chorion has been completed, the eggshell is still quite permeable to water. It is not until later that there is a further secretion of lipid material, probably by the oöcyte itself, which passes into the space between the vitelline membrane and the inner layers of the chorion. At this stage the permeability decreases markedly, indicating that a layer of orientated lipid molecules has formed over the surface of the egg membranes. The permeability of this layer is affected by temperature in much the same way as is that of epicuticular wax layers, and this relation will be discussed further in Chapter 16.

The formation of the egg membranes completes the process of oögenesis, and the egg ruptures its follicle to pass into the oviduct for fertilization and eventual oviposition. The form in which eggs are deposited varies greatly from species to

Fig. 12.2. Oögenesis in insects. (a) Schematic illustration of the three main types of ovariole. A, panoistic type without nurse cells; B, polytrophic type with nurse cells interposed between successive oöcytes; C, telotrophic type with nurse cells confined to the apex (de Wilde, 1964). (b) Oögenesis in the polytrophic ovary of *Anopleura*. A, stage showing six undifferentiated oögonia; B, the six cells differentiated into five nurse cells and one oöcyte; C, the enlarged nucleus of the oöcyte giving off chromidia to the plasma; nurse cells beginning to function, next group of oögonia visible above; D, nuclei of follicular cells enlarge; nurse cells at height of activity; plasma with numerous chromidia; E, follicular epithelium active; nurse cells beginning to regress; chromidia formation ceases; fat droplets appear in plasma; F, follicular epithelium at height of activity; nucleus of oöcyte dissolves, chromosomes free in plasma at periphery of cell; yolk formation begins; G, ripe egg with reserve materials complete; follicle cells have laid down chorion; spindle of first maturation division formed (Wigglesworth, 1965 after Ries).

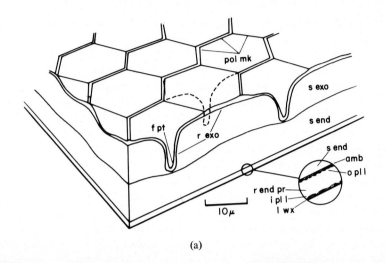

(a)

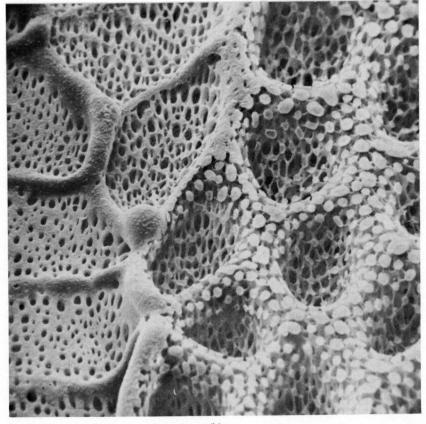

(b)

species; they are usually coated with secretions of the accessory glands, the details of whose activity have been investigated in a number of cases. In the cockroach quite different types of secretion are produced by the right and left members of the paired collaterial glands. The left gland, itself differentiated to form distinct regions, secretes a highly viscous liquid containing the protein which will form the structural basis of the oötheca, a β-glucoside of protecatechuic acid and a polyphenol oxidase. The secretions of the right gland contain a β-glucosidase, so that when the secretions mix in the oviduct, protecatechuic acid is liberated to serve as the substrate for the oxidase, and the quinones so produced form cross-linkages with the structural protein to give the hardened and darkened substance of the egg case (Brunet and Kent, 1955).

In mantids, the oötheca is formed from a viscous secretion of the accessory glands, which is beaten into a frothy mass that hardens on exposure to air and forms a vacuolated nest in which the contained eggs undergo embryonic development. It is uncertain to what extent the provision of these different kinds of egg pod is significant in relation to the protection of eggs from the desiccating influence of the terrestrial atmosphere. They are absent in many species which lay their eggs singly, and it is likely that the egg membranes themselves provide adequate water-proofing, and that the significance of the elaborate egg rafts formed in many species is to be sought in quite other terms.

b. Spermatogenesis

The testes comprise a number of tubular follicles, which contain the germ cells in different stages of development (Fig. 12.4(a)). At the apex of the follicle lie clusters of germ cells, or spermatogonia, interspersed with somatic nurse cells; as the spermatogonia move down the tubule they become invested in a layer of somatic cells, which form a cyst within which the germ cells undergo successive divisions to form the haploid spermatids. At the proximal end of the tubule, the spermatids undergo transformation to form spermatozoa, a process which involves the concentration of nuclear material to form the head, and the development of the characteristic flagellum. At this stage the spermatozoa break through the walls of their cyst and migrate to the seminal vesicles (see Fig. 12.1(b)), where they become densely packed to form a reservoir of mature sperm.

Fig. 12.3. Membranes of the insect egg. (a) The chorion of *Rhodnius*. Note the polygonal markings of the shell surface (pol mk), each with a follicular pit at its centre (f pt). The resistant exochorion layer is thickened at the base of each pit and at the surface ridges (r exo); s exo, soft exochorion layer; s end, soft endochorion layer; *Inset:* the detailed structure of the resistant endochorion layers; amb, amber layer; i pll, inner polyphenol layer; o pll, outer polyphenol layer; r end pr, resistant endochorion protein layer; l wx, primary wax layer (Beament, 1946). (b) Plastron of part of the egg shell of a fly, *Fannia armata* (Meig.). Air is held in a widely spaced meshwork as well as in a lower and finer meshwork, which shows high resistance to hydrostatic pressure. From unpublished scanning electron micrograph (× 1550) courtesy of Professor H. E. Hinton.

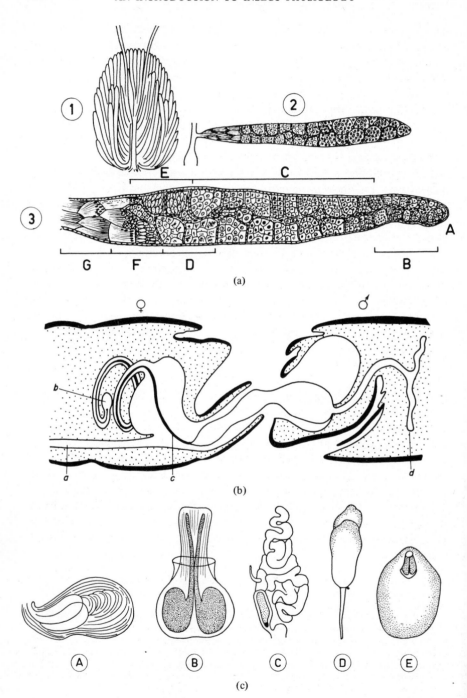

(a)

(b)

(c)

c. Impregnation

The external genitalia of insects are enormousy complex, and the wealth of interspecific variation in structural detail has served as an invaluable basis for the taxonomy of the group. It seems likely that the difficulty of effecting a tight coupling between male and female genital tracts, the indispensable prerequisite for efficient internal fertilization in a terrestrial environment, must have posed a formidable problem for animals of the size of insects, and it is this that finds reflection in the intricate and highly specific configuration of their copulatory appendages. The "key-and-lock" principle suggested by their anatomical elaboration was originally thought to be of special significance in relation to the prevention of interspecific mating, but now it seems more likely that the main barrier to such mating occurs at the level of precopulatory courtship. The close and detailed fit of male and female genitalia should probably be interpreted rather in terms of the effective coupling which it ensures between the sexes during the period of sperm transfer, a process that may occupy several hours. In some species the sexes may actually become cemented together during copulation by special secretions of the male accessory glands.

The insertion of the penis into the female's reproductive tract appears to be effected by a hydraulic mechanism, fluid being pumped into the phallus by compression of the abdomen and, in some species, by the activity of a fluid pump situated at the posterior end of the ejaculatory duct (Fig. 12.4(b)); as erection proceeds, the penis gradually penetrates through the spermathecal duct to the spermatheca itself, into which the sperm is ejaculated.

In many insects the spermatozoa are introduced into the genital tract of the female as free suspensions in a seminal fluid. This fluid is elaborated in part by the accessory glands, and contains a high concentration of free amino acids, protein and carbohydrate. In others, the semen is enclosed in a membranous proteinaceous sheath produced, in the form of a compact spermatophore, from secretions of the accessory glands, often under the influence of other parts of the genital tract. This spermatophore is sometimes inserted by the male into the spermatheca of the female, but more usually it is deposited in the vagina, or in some cases simply dropped by the male during courtship, to be picked up by the female for deposition in the vagina. The precise form of spermatophores varies greatly from species to species (Fig. 12.4(c)), and their shape is often related

Fig. 12.4. Aspects of the male reproductive system. (a) Testis and follicle of a grasshopper. 1, general configuration of the testis; 2, single follicle with vas efferens; 3, enlarged view of follicle. A, apical cell; B, spermatogonia; C, spermatocytes; D, first maturation division; E, second maturation division; F, spermatids; G, spermatozoa (de Wilde, 1964 after Schröder). (b) Copulation in *Lygaeus equestris*, showing extension of penis of the male into the receptaculum seminis of the female (schematic); a, oviduct; b, receptaculum seminis; c, penis; d, vesicle containing fluid which is driven into the penis to extend and uncoil it (Wigglesworth, 1965 after Ludwig). (c) The spermatophores of different species of insect. A, *Blatella germanica*; B, *Sialis lutaria*; C, *Anabolia nervosa*; D, *Galleria mellonella*; E, *Pimpla instigator* (de Wilde, 1964 after Khalifa).

quite precisely to the shape of the female's vaginal chamber. They may be provided with apertures or tubes through which, after insemination, the spermatozoa can escape, to make their way to the spermatheca of the female. Mechanical pressure exerted by muscular contractions of the female's reproductive tract may contribute to the emptying of spermatophores, but in many insects it is only the neck of the spermatophore that is inserted into the genital tract of the female, the body remaining outside; here it seems probable that swelling of a gelatinous component of the spermatophore may serve to push out the contained seminal fluid. In some insects, the empty spermatophore case is eaten by the female, in others it appears to be digested by proteinases of the genital tract, and absorbed through its walls.

The mechanism by which spermatozoa, that have been deposited in the vagina, get to their eventual destination in the spermatheca appears to differ in different species. In some the sperm seem to move actively towards the spermatheca, possibly in response to a chemical stimulus; in others they appear to be transported passively by peristaltic contractions of the genital ducts. In *Rhodnius* it has been shown that such contractions can be induced by a component, possibly an *o*-dihydroxy-indolalkylamine, of the secretions of the male accessory glands.

d. Fertilization

Following impregnation, the spermatozoa are stored in the spermathecae until such time as they are required for the fertilization of eggs descending from the ovaries. The females of many insects mate only once, and the sperm from that first mating suffice to fertilize eggs during the whole of the egg-laying period, which may last for months and even years. The recruitment of sperm for fertilization appears, in some species, to be effected by the contraction of special muscles associated with the spermathecae, which pump out a batch of spermatozoa at the appropriate time. Once the spermatozoa have entered the oviduct near the micropyle of the egg, they may be guided to the micropyle by chemical stimuli; as soon as they have penetrated through the wax layer of the egg, a fertilization membrane is deposited between the wax layer and the oöcyte, and the spermatozoan loses its tail to transform into the male pronucleus. This combines with the female pronucleus to form the zygote, and thus the process of development is initiated.

CHAPTER 13

DEVELOPMENT

The development of insects, as of other animals, from embryo to adult, involves two quite different processes, which can conveniently be discussed separately, though they take place concurrently; they are growth, which is referable to an increase in the number or size of cells, and differentiation, which involves a change in the pattern of their metabolic activity.

1. Growth

Growth is a readily definable phenomenon, which can often be measured in terms of cell number, or more simply in terms of size or weight. Since adult insects are small, the amount of growth that occurs during development from the egg is relatively slight, but since insects possess an external skeleton, it is much more markedly cyclical than it is in most other animals. The possession of a rigid cuticle effectively limits the size of an insect, and it can only grow in size by discarding its old exoskeleton and making a bigger one. Increase in size, therefore, tends to occur as a stepwise, discontinuous process; a phase of cell multiplication is followed by ecdysis, the moulting of the old cuticle, and an increase in size, with the deposition of a new cuticle conformable to the new size. So there is a succession of long intermoult periods, during which there is no change in linear dimension, alternating with short moults, during which a sharp increase in linear dimension occurs (see Fig. 13.1). If weight, rather than a linear measurement, is used as a measure of size, the discontinuities of the growth process are of a different kind; little increase in weight takes place during the short ecdysial phase, while the intermoult period is characterized by a substantial increase, reflecting the deposition of food reserves, and an increase in the size of cells (Fig. 13.1).

The number of moults which occur during development, the number of instars interposed between egg and adult, varies greatly from species to species. At one extreme is an insect like the mealworm beetle, which may have as few as four instars; at the other the firebrat, *Thermobia,* where the number of instars is indefinite as moulting continues in the adult, and as many as 60 may be passed

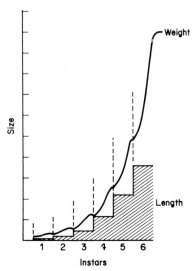

Fig. 13.1. The increase in size during growth; the hatched (lower) curve shows the increase in linear dimensions of the mealworm beetle; the upper curve shows the increase in weight of the stick insect, during successive instars. (Redrawn from Wigglesworth, 1965 after Teissier.)

through before death. There may be considerable variation in the number of instars within a given species, depending on genetic factors such as sex, or on external factors such as temperature or level of nutrition.

The succession, in the life cycle of insects, of a series of instars separated by ecdyses, imposes a cyclical pattern on the activity of epidermal cells. This involves the resorption of parts of the old cuticle, leading to a weakening of its structure, in preparation for the act of moulting; usually an increase in cell number, and an expansion of the epidermal surface following ecdysis; and the deposition by the epidermal cells of a new exoskeleton. These cyclical activities of the epidermis form the basis for the whole process of growth in insects, and consideration must be given to the details involved.

The structure of the cuticle has been briefly outlined in Chapter 1; it is illustrated again in Fig. 13.2, together with the cellular elements which contribute to its formation. These include:

(a) the epidermal cells;

(b) the oenocytes, characterized by massive accumulations of lipoprotein at the time of moulting;

(c) the dermal glands, also characterized by lipoprotein deposits, discharging to the surface of the cuticle through special ducts; and

(d) the tormogen and trichogen cells, which are responsible for the formation of sensory bristles, together with the neurones which innervate the sense organs.

All of these cell types have a common origin, arising by differentiation from cells of the epidermis; the last type of cell associated with the cuticle is mesodermal in origin; it is

(e) the haemocyte, which appears to be mainly responsible for the laying down of a basement membrane, that serves to support the epidermal elements. At the time of deposition of this membrane, when epidermal cells have completed their multiplication, the material contained in the haemocytes appears to be discharged to form a component of the substance of the basement membrane.

In the nymphal instars of *Rhodnius*, the cycle of epidermal activity is triggered by the act of feeding, and the sequence of events is illustrated in Fig. 13.2(b). Prior to the taking of a blood meal, the epidermis is in a relatively inactive state, but activation occurs soon after feeding. It involves a conspicuous enlargement of the nucleoli and the appearance of high concentrations of ribonucleoprotein in the cytoplasm, indicating that the cells are entering upon a phase of protein synthesis. On about the fifth day after feeding, the epithelial cells begin to divide, and a phase of intense mitotic activity follows, leading to an increase in the density of epidermal cells. At the same time, the conformation of the epithelial cells changes from squamous or cuboidal to columnar. At this stage the epidermal cells separate from the cuticle, and begin to lay down a new cuticle; they also secrete, into the space between the old and the new cuticle, a moulting fluid containing proteolytic enzymes and chitinase. Under the influence of these enzymes, the old cuticle begins to be digested away. The new cuticle is unaffected, because it is protected by the layer of cuticulin which is the first to be deposited, appearing as a delicate membrane that covers the folded surface of the epithelial cells, thus providing a capability for expansion once the old cuticle has been shed. The deposition of cuticulin is associated with a decrease in size of the greatly swollen oenocytes, and it is reasonable to suppose that the secretions of these cells furnish a raw material for the formation of the cuticulin layer.

Below the cuticulin is deposited the "homogeneous inner epicuticle" composed, as the cuticulin layer probably is, of a tanned lipoprotein complex. Below these there are in turn deposited the successive lamellae of the procuticle, at the same time that the definitive structure of the epicuticle is completed. Continuity between the epicuticle and the epidermis is maintained by the existence of pore canals, which in most cuticles traverse the lamellae of the protocuticle to terminate at the boundary between pro- and epi-cuticle. Through these pore canals, the epidermal cells secrete a material rich in polyphenols, that contributes to the formation of the epicuticle. They also constitute the route by which, just before moulting, waxes are secreted to ensure the water-proofing of the new cuticle, shortly to be exposed to the atmosphere.

The cuticular waxes are secreted as long, filamentous structures, 60—130 Å in

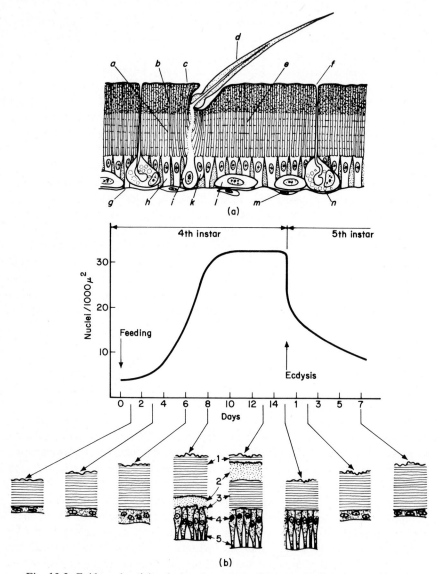

Fig. 13.2. Epidermal activity during the growth of insects. (a) Schematized diagram of the epidermis and associated structures. a, laminated endocuticle; b, exocuticle; c, epicuticle; d, bristle; e, pore canals; f, duct of dermal gland; g, basement membrane; h, epidermal cell; i, trichogen cell; k, tormogen cell; l, oenocyte; m, haemocyte adherent to basement membrane; n, dermal gland. The sense cell and axon of the bristle have been omitted. (Wigglesworth, 1965.) (b) Schematized diagram showing changes in the density of epidermal cells during the moulting cycle, associated with the resorption of old and the deposition of new cuticle in fourth and fifth instars of *Rhodnius*. 1, cuticle of fourth instar; 2, moulting fluid; 3, cuticle of fifth instar; 4, epidermal cells; 5, basement membrane (redrawn from Wigglesworth, 1959 and Locke, 1964).

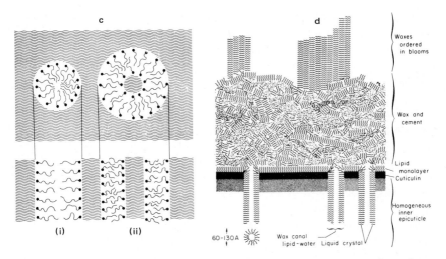

Fig. 13.2. (cont.) Lipids of the insect cuticle. (c) Diagram of the structure of two phases of lipid/water liquid crystals; above, transverse and below, longitudinal sections; (i) the middle phase; (ii) the complex hexagonal phase (Locke, 1964 after Luzzati and Husson). (d) The structure of the surface of an insect. Not all insects have all the illustrated components; the cement is frequently lacking, and wax blooms would then form directly over the wax canals, or the waxes may be liquid and mobile, in which case blooms would not form (Locke, 1964).

diameter, and visible in electron micrographs. These filaments appear to represent lipid/water liquid crystals, which may exist in a number of different phases (see Fig. 13.2(c)). The "middle phase" crystals have the hydrophobe carbon chains directed towards the interior of the filament, with the hydrophil, polar groups towards the lipid/water interface at the surface of the filament. In the "complex hexagonal phase", the filaments are double-walled, with hydrophil polar groups both inside and outside the filament. These liquid crystals appear to be discharged from the tips of the pore canals, and to penetrate the epicuticle through minute wax canals that extend through the cuticulin layer to the surface. It is envisaged that the middle phase crystals are continuous with a surface monolayer of lipid (see Fig. 13.2(d)), orientated with polar groups towards the hydrophil polyphenol substratum of the cuticulin layer. Lipids which cannot be accommodated in the monolayer are extruded to its outer surface, where they form a less well-orientated layer, mixing with the cement which is being secreted from the dermal glands in a complex which may serve to protect the monolayer. In many species of insect wax-blooms, composed of thin sheets, or plates, of orientated lipid molecules may project from the general surface, as shown in the diagram.

This recent work has provided a satisfactory explanation of a phenomenon which has long puzzled insect physiologists, namely the secretion of insoluble lipid material through a water-impregnated cuticle. It should be mentioned, however, that it does so on the basis of the participation of lipid molecules which have one end polar and hydrophil, the other non-polar and hydrofuge. Only a small proportion of the cuticular lipids of insects are, in fact, of this type. It would presumably be possible to accommodate non-polar lipids within the hydrophobe core of a middle-phase crystal, but evidence for such an arrangement is lacking.

The mechanism by which the lamellae of the procuticle are deposited below the epicuticle has not yet been unequivocally established, but it appears to involve the secretion of cuticular material (chitin and protein) from the base of the finger-like processes that project into the pore canals from the epidermal cells. The lamellar structure seems to be associated with a rhythmic activity of the epidermal cells, most of the material being deposited during the night. Only that part of the procuticle which is destined to become tanned exocuticle is laid down at the time of moulting, the soft endocuticle continuing to form during the intermoult period, right up to the time of the next moult.

During the time when the epicuticle and presumptive exocuticle are laid down, the bulk of the old cuticle is being digested away by enzymes of the moulting fluid, and the fluid plus dissolved digestion products are resorbed from the space between the old and the new cuticle. Precisely how this is achieved has not yet been determined, but it is possible that absorption takes place through the wax and pore canals.

As the insect emerges from its old, attenuated cuticle, which splits along ecdysial lines of weakness, it swallows air to expand its bulk. The swallowing of air, together with tonic contractions of the muscles of the body wall, ensures the development of substantial haemolymph pressure, and under the influence of this pressure, the soft and extensible new cuticle stretches to take up the conformation of the succeeding instar, thereby reducing the density of epidermal cells as illustrated in Fig. 13.2(b). Once this has occurred, the cuticle begins to darken and harden by the mechanism outlined in Chapter 1. The polyphenol tanning agents appear to be discharged from the tips of the pore canals, and to diffuse inwards, so that darkening and hardening occurs first in the cuticle which adjoins the epicuticle, proceeding inwards from there.

Growth of the epidermis is thus seen to be a particularly complex phenomenon, because it is restricted by the presence of a rigid exoskeleton. It involves the partial dissolution of that skeleton, and the deposition of a new one, once expansion of the epidermal surface, based on an increase in the number of epidermal cells, has occurred. In tissues other than the epidermis and its derivatives, this complication does not arise, and here all that need be involved is a multiplication or growth of cells, and their differentiation along lines appropriate to the particular tissue concerned. It should be emphasized, however, that the growth of tissues is not based on a simple process of cell division. In many insects, there may be little or no cell multiplication during larval instars, and increases in the surface area of the epidermis are associated with an increase in the size, rather than the number, of epidermal cells. Where multiplication of cells does occur, the rate of cell division during periods of growth is generally excessive, so that many more cells are formed than are required. The surplus of cells undergoes histolysis, the nuclei disintegrate to form "chromatin droplets", and the products of histolysis are either assimilated by surviving sister cells, or they diffuse into the haemolymph to become part of the general metabolic pool, while cellular debris may be cleaned up by phagocytes of the haemolymph (see Chapter 3). At any moment, therefore, the number of cells present reflects the state of balance between mitotic activity on the one hand, and cell destruction on the other; during later stages of the growth cycle, histolytic processes may outstrip mitotic recruitment, so that the number of cells actually decreases.

2. Differentiation

In insects, as in most other animals, growth in size is usually associated with a change in form. One instar does not differ from the next simply in that it is bigger; at the least there will be some change in the relative proportion of parts, and often profound differences in the details of structural organization are involved, a differentiation of one instar in relation to its predecessor. The degree to which such differentiation occurs at moulting to the adult forms the basis of a subdivision of the class into three main groups. In the Ametabola the change from young to adult is a gradual one, achieved in the course of a succession of nymphal instars. The point at which these insects become sexually mature does not necessarily mark the cessation of growth and moulting, nor is it associated with a change in form which could rightly be described as metamorphic. Insects in which active nymphal stages are transformed directly into the adult form without the intervention of a quiescent pupal stage are classed as Hemimetabola; here the change in form at the last moult to the sexually mature adult is usually substantial in comparison with the change between successive nymphal instars. Species where a comparatively inactive non-feeding pupal stage is interposed between the immature forms, normally referred to as larvae, and the sexually mature adult are included in the Holometabola, showing complete metamorphosis. What is ultimately distinguished in this classification is the magnitude of the metamorphic change that occurs when the insect moults to the sexually mature form; the change is small in the Ametabola, great in the Holometabola and intermediate in the Hemimetabola. The difference is clearly one of degree, and it has been found to reflect a corresponding difference at the level of developmental physiology, to which attention must in the first place be directed.

The earliest stages in the development of insects are characteristically regulative, in the sense that products of the first few cell divisions are totipotent, each capable of giving rise to a fully-developed organism if other division products are destroyed or eliminated by microsurgical techniques. As development proceeds, the embryo becomes progressively more of a mosaic, so that extirpation of cells in one part of the embryo will result in the development of individuals with deficiencies in the corresponding region. The different cells, exposed to different influences as a result of the developmental process, become determined to differentiate along certain lines, and are therefore no longer capable of substituting for cells determined to a different fate. The process of determination involves the suppression, often irreversible, of certain potentialities, and the manifestation of others, out of the total represented in the genotype. Initially, for example, all cells would have the prospective capability for the production both of muscle proteins and of cuticular proteins; but in the cells which are, in the course of development, induced to form muscle cells, the capacity to produce cuticular proteins is suppressed or lost, while in prospective epidermal cells, the capacity to produce muscle proteins is suppressed or lost. Each type of cell, by emphasis on this or that metabolic pathway, develops its own particular pattern of metabolism, ensuring the production of this or that particular reaction product, a structural protein, a hormone or a digestive enzyme, appropriate to the performance of its function in the organism as a whole. In the course of embryogenesis, there will be formed a corresponding diversity of tissues, serving as the basis of corresponding organ systems.

The post-embryonic development of the Ametabola can be seen as a simple extension of such a process of progressive differentiation; it involves a gradual increase in the size of the insect, with perhaps some slight changes in the relative proportion of its parts. This would be capable of being achieved without a change in the nature of the activity of particular types of cells; the epidermal cells would continue to secrete the same type of cuticle, the muscle cells the same type of muscle protein at all stages of development. Only the sex cells come under what could be described as a metamorphic influence; during early stages their differentiation is completely suppressed, but as the insect nears the completion of development, the suppression is lifted, and the cells proceed to differentiate to their functional stage in the sexually mature insect.

In the Hemimetabola the situation as far as the sex cells are concerned is much the same, but here the last moult to the sexually mature adult is associated with a substantial change in the general form of the insect. *Rhodnius* is a member of this group whose development has been particularly carefully investigated. In this insect the metamorphic moult is characterized, among other things, by the formation of large membranous wings in place of the inconspicuous wing lobes of earlier instars, and the development of external genitalia. These formations express the morphogenetic movements of corresponding regions of the epidermis, but even in parts which are unaffected by such changes in surface conformation there is a marked difference in the type of cuticle laid down before and after the metamorphic moult. The nymphal cuticle is thick, soft and extensible, and the surface is beset with small plaques that bear the sensory bristles (see Fig. 13.3(a)). The adult cuticle, on the other hand, has fewer bristles of a different type, and is thin and heavily sclerotized; distension of the abdomen during feeding is provided for not by extensibility of the general cuticle, but by the unfolding of lateral pleats of soft cuticle. The pattern of pigmentation also differs considerably, with nymphal pigment spots confined to the postero-lateral margins of the abdominal sclerites, while the adults have larger spots at the antero-lateral margins. This furnishes a good example of the way in which profound changes in the general appearance of an insect can be based largely on changes in the activity of epidermal cells. From the point of view of developmental biology the phenomenon may be said to represent a retention of pleuripotency on the part of the epidermal cells. The cells have not been irrevocably determined to the laying down of a particular type of cuticle, but have retained a capacity to lay down two different types, so that the same cell that at one moult produces cuticle of one sort, at the next lays down quite a different sort; a potentiality which was suppressed at the first moult finds expression in the next.

That the capacity to lay down nymphal and adult cuticle is present latently in the same cell is well illustrated by experiments involving injury to the epidermis. If a section of cuticle is injured by burning, for instance, the uninjured cells at the margin of the wound divide and spread inwards to repair the damage, and as they do so they carry with them their special characteristics. If the burn passes near a black pigment spot, the proliferation and inward spread of the corresponding cells is reflected, at the next moult, in a corresponding extension of the pigment spot, as shown in Fig. 13.3(b); or if one pigment spot is burnt out, its place at the next nymphal moult will be taken by cells which have spread from adjacent non-pigmented areas. But if the following moult is to the adult, then the situation changes, for the pigment spots of the adult occupy places which in the nymphs are unpigmented, and vice versa. Hence, where the nymph had an extended pigment spot, the adult will have unpigmented cuticle in place of its normal pigmentation, and where the nymph lacked a pigment spot, there the adult will have an extra pigmented area.

Such experiments show that the capacity to produce the adult pattern is present in the nymphal cells, and if they are induced to divide at the nymphal stage, they distribute their potentiality to their daughter cells; but the adult pattern remains latent, to find expression when metamorphosis occurs. If this is so, one would expect that, provided "metamorphic" conditions could be provided, nymphal cells might exhibit adult characteristics at any stage in nymphal development, and conversely, that adult cells might be induced to exhibit characteristics of the nymphal cells, provided they could be subjected to a nymphal environment. Within limits this can, in fact, be shown to be the case.

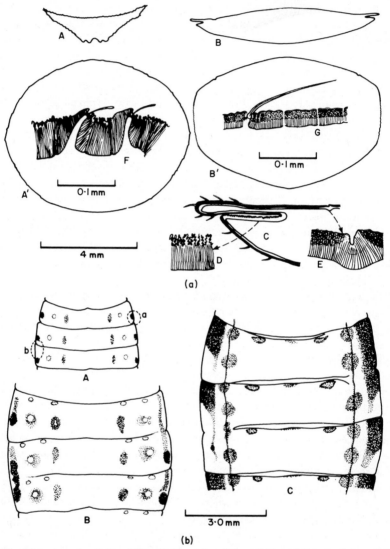

Fig. 13.3. Aspects of the differentiation of cuticle. (a) Nymphal and adult cuticle of *Rhodnius*. A, transverse section of abdomen of unfed fifth-stage nymph; A′, the same immediately after feeding. B, transverse section of abdomen of unfed adult; B′, the same immediately after feeding. C, detail of lateral pleat in abdomen of unfed adult; D, detail of extensible lower wall of the pleat; E, detail of "hinge-line" in tergites. F, longitudinal section of abdominal tergite of fifth-stage nymph; G, longitudinal section of abdominal tergite of adult (Wigglesworth, 1959). (b) The effect of injury on the development of pigment spots in nymphal and adult *Rhodnius*. A, third, fourth and fifth tergites of a normal third-stage nymph; the broken lines at "a" and "b" show the regions burned. B, corresponding segments in the fifth-stage nymph resulting. C, corresponding segments in the adult resulting (Wigglesworth, 1959).

If, for example, the endocrine organs responsible for the maintenance of the nymphal state (see next chapter) are implanted below the cuticle of a fourth instar nymph, then when the insect moults to the adult form, the cuticle covering the implant will be of the nymphal type, surrounded on all sides by normal adult cuticle.

While the metamorphic changes associated with the activity of the epidermis are the most spectacular, and have been the most carefully studied, they are by no means the only ones involved in the metamorphic moult. The formation or elaboration of movable appendages, like wings and external genitalia, will clearly be associated with corresponding changes in musculature, which may involve both degenerative and regenerative processes. Other internal organs, like the tracheal system, the nervous system or the excretory system tend, by contrast, to be relatively little affected by the metamorphic moult.

While the metamorphosis of the Hemimetabola may be sufficiently spectacular, as in the transformation of a virtually wingless young to the fully-fledged adult, the changes involved are trivial compared with the total transformation that occurs during the metamorphosis of holometabolous species. In them the immature insect may be a soft, white and virtually featureless maggot, anatomically and physiologically adapted to a simple mode of life; this larva will develop, through the pupal stage, to its winged and six-legged, firmly sclerotized and finely sculptured adult counterpart, quite another animal with a different and much more complex mode of life. The change from larva to adult could in this case be thought of as being too far-reaching to be accomplished in a single moult, hence the interposition of a quiescent pupal stage, devoted to the cellular reorganization involved; and too drastic to find a basis in the capacity of epidermal cells to differentiate first along immature and later along adult lines. Instead, the task of adult reconstruction has in this group been delegated to special embryonic cells, whose development is totally suppressed during larval life. Clusters of such cells form the imaginal buds, or histoblasts, which can be found dispersed among the larval tissues, most conspicuously in regions where adult appendages are destined to make their appearance as antennae, mouthparts, legs, wings and external genitalia. There is thus no question of a change in the direction of development of epidermal cells from larval to adult patterns, but rather of an almost complete replacement of larval epidermal cells by cells produced from imaginal buds, and differentiating along adult lines. In these insects, too, the striking difference between larval and adult modes of life render many of the internal organs of the larva unsuitable to perform their function in the context of the adult organization, and a need arises for the partial or total replacement of musculature, alimentary canal, salivary glands, fat body etc. The pupal stage is therefore a period of intense histolytic activity, which furnishes the raw materials for the histogenesis of imaginal buds.

Three main types of developmental process seem thus to be involved in the mechanism by which the adult stage is attained in insects. One is a progressive differentiation of somatic cells and tissues towards the adult condition. Another involves the retention of embryonic characteristics, mainly by cells of the epidermis, which enables a switch to be made in the pattern of development at the time of metamorphosis, so that cells which had previously disposed themselves to form the nymphal surface, and secreted nymphal cuticle, now become the source of morphogenetic movements that tend to the formation of the differently disposed adult surface, covered by the adult type of cuticle. In the third type, two sets of cells are formed during early stages of development, one of which is responsible for the formation of the larval organization, while the other is held in reserve, its differentiation suppressed during larval life; activity is resumed in these cells some time before the metamorphic moult, and on the basis of their growth and differentiation the adult organism is formed to replace the crumbling fabric of larval tissues. As has already been indicated, these three processes are involved to different extents in the development of the three groups of insects, the first serving as the main basis of ametabolan, the second of hemimetabolan and the third of holometabolan development. It must be emphasized, however, that the distinction that can be drawn in these terms is again one of degree; histoblasts, for instance, are a feature of the development in many Hemimetabola as well as of the Holometabola, while epidermal pleuri-potency is involved in the metamorphosis of a number of Holometabola as well as in that of the Hemimetabola. A capacity to manifest all three types of developmental process appears to be shared by all members of the class, but the emphasis is shifted from one to another in accord with the demands made in terms of reconstruction at the metamorphic moult. The feature that is common to all is a suppression of adult characteristics during early stages of development, and the suppression of larval characteristics at metamorphosis, and the nature of the control systems responsible for this change will be discussed in the next chapter.

3. Diapause

One aspect of the growth and development of insects that must be briefly mentioned is the phenomenon of diapause, or more generally, of arrested development, which is an important feature of the life history of many species of insects. It appears to be essentially a mechanism for limiting the occurrence of delicate morphogenetic processes to periods during which environmental conditions are favourable; or, alternatively, for synchronizing the life cycle with seasonal fluctuations of climate in such a way as to ensure that an abundance of food will be available for active stages of the life history.

In many species, arrest of development, or quiescence, may occur as a direct result of inclement conditions; poor food, for instance; or drought and desiccation; or excess moisture; or low temperatures. In others, however, a prolonged arrest of development may occur, at one or another stage of development, quite irrespective of the current environmental situation. The timing of diapause in such species is still linked to seasonal changes, but more remotely, and any one of a number of different environmental factors may be involved in the induction of diapause. Of these, photoperiod appears to be one of the most important, and Fig. 13.4(a) illustrates the relation between day-length and the intensity of diapause in different species, as measured by the proportion of individuals that exhibit developmental arrest. In one of the species, larval diapause is induced over a narrow range of day-lengths between 10 and 14 hr; in another, exposure of eggs to day-lengths in excess of 16 hr causes the production of a high proportion of diapause eggs; and in another the proportion of diapause pupae increases gradually as the period of daylight increases from 0 to 12 hr, and then falls again to zero.

Diapause in temperate species of insect can usually be seen as a mechanism for surviving during the winter season, when temperatures drop to levels that are too low for normal developmental processes. It is not surprising, therefore, that exposure to low temperatures is one of the best ways of breaking the diapause. The mechanism of this effect seems to be based on the fact that the corresponding physiological process, which may be different in different insects but which leads to what may be called the breaking of diapause, occurs at high rates only when the temperature is low. This is illustrated in Fig. 13.4(b) for the embryonic development of an Australian grasshopper. Curves on the left show the proportion of eggs in which the breaking of diapause is completed in 60 days, as determined on samples originating in different parts of the country. It is highest at temperatures between $10°$ and $15°$, falling to lower values at lower temperatures, and to zero at about $25°$. At temperatures which ensure rapid breaking of diapause, however, the rate of normal morphogenesis is negligible, as shown by curve B. To effect rapid development, eggs must therefore be exposed first to low temperatures, which will hasten diapause development, and then to high temperatures, which will allow morphogenesis to proceed rapidly. The phenomenon of diapause seems, in other words, to be based on an inhibition of some part of the developmental process, an inhibition which can only be released by a reaction that has a temperature coefficient different from, and usually lower than, that of morphogenesis. Where the two temperature curves show a degree of overlap, as in the illustration, development will proceed slowly to completion at a constant intermediate temperature; where there is no overlap, as is the case with eggs of the silkworm, development cannot occur at intermediate temperatures; and where the optima are so close together that overlap is extensive the existence of diapause may be difficult to establish experimentally.

The environmental factors which induce and break diapause have been

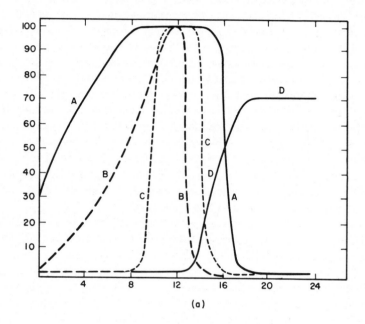

(a)

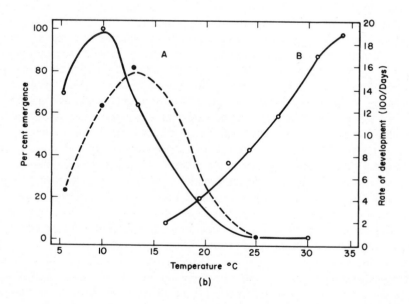

(b)

identified for a number of different species of insect, but the mechanism by which they exert their effect is still in doubt. It seems to be generally agreed that the immediate cause of diapause is lack of growth hormone, a substance whose source and properties will be described in the next chapter; but precisely how the arrest of neuroendocrine secretion is ensured has not yet been established.

Fig. 13.4. The relation between diapause and certain environmental factors. (a) The effect of photoperiod on the incidence of diapause in some species of Lepidoptera. Ordinate: percentage of individuals entering diapause; abscissa: hours of light in 24 hr. A, *Acronycta rumicis* (after Danilyevsky); B, *Grapholitha molesta* (after Dickson); C, *Pyrausta nubilalis* (after Beck); D, *Bombyx mori,* bivoltine race (after Kogure) (Wigglesworth, 1965). (b) The influence of temperature on the breaking of diapause and on morphogenesis in the embryo of *Austroictes cruciata.* A, the ordinate shows the proportion of eggs to complete the breaking of diapause during 60 days at the specified temperatures; complete line: eggs from South Australia; broken line: eggs from Western Australia. B, the ordinate shows the proportion of post-diapause development completed each day at the stated temperatures (Andrewartha and Birch, 1954).

CHAPTER 14

NEUROENDOCRINE CONTROL SYSTEMS

Brief mention has been made in earlier chapters of the control by endocrine secretions of somatic processes like digestion and excretion, and of neuro-muscular processes like activity rhythms. It seems likely that as more of the details of such physiological processes become known, so will the study of their neuroendocrine control systems become of increasing importance. It is in the fields of developmental and reproductive physiology, however, that neuro-endocrine regulators play their most spectacular role, and it seems appropriate, therefore, to have delayed a detailed discussion of the systems involved till now, when the main features of reproductive physiology have been outlined. The control of growth and metamorphosis by the neuroendocrine system of insects has, indeed, provided one of the most intriguing problems for insect physiologists since the discovery by Kopeč, in the early decades of this century, that the moulting of insects is under the control of a circulating hormone. A great deal of outstanding experimental work has been done to elucidate the factors involved, and on the basis of results obtained it has been possible to formulate a satisfactory interpretation of the general situation, though much remains to be done at the level of specific detail.

1. Neuroendocrine Control of Growth and Development

In order to provide a broad framework for the discussion that follows, it will be useful to review briefly the main features of the neuroendocrine system concerned, and to outline its mode of action in general terms. The components of the system are illustrated diagrammatically in Fig. 14.1, and they are seen to comprise four main parts:

(a) clusters of neurosecretory cells in the brain, whose secretions are passed along nerve axons to

(b) the corpora cardiaca, constituting a neurohaemal organ associated with the dorsal blood vessel; through this organ the brain secretions are passed into the bloodstream to exert an activating influence on

(c) the thoracic glands; these glands then secrete a hormone called the

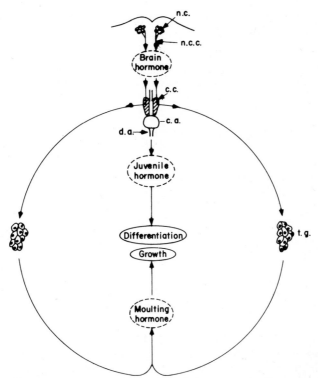

Fig. 14.1. A general outline of components of the neuroendocrine system concerned with the control of growth. c.a., corpus allatum; c.c., corpora cardiaca; d.a., dorsal aorta; n.c., neurosecretory cells of the protocerebrum; n.c.c., nervi corpora cardiaca; t.g., thoracic glands.

moulting hormone, or ecdysone, which initiates the cycle of growth that has been described in the previous chapter;

(d) the last major component of the complex is the corpus allatum, closely associated with the corpora cardiaca. The activity of this gland is under the control of the brain, and it produces a hormone known as the juvenile hormone, which is passed into the bloodstream; when the hormone is present in sufficiently high concentration at the time of moulting, juvenile, or larval, characters are manifested, while in its absence, the adult characteristics make their appearance.

Against this background, the general growth and development of insects may be seen as governed by the cyclical activity of the neurosecretory cells of the brain. The hormone produced activates the thoracic gland, which secretes moulting hormone to induce a corresponding succession of moults. During early stages, the activity of the corpus allatum ensures the appearance of juvenile characters; with decreasing activity and a corresponding decline in the titre of juvenile hormone, pupal, and eventually adult, characters supervene. On the basis of this interpretation, interest will centre on the nature of substances released by the various glands, the nature of their action, and the mechanisms by which the secretory activity is triggered to produce co-ordinated cyclical activity in the system as a whole; these aspects will be discussed in the sections that follow.

a. The Neurosecretory Cells and the Corpora Cardiaca

The neurosecretory cells are situated in the median dorsal region of the brain; at certain points in the cycle of their activity, the cell bodies contain an accumulation of secretory material in the form of minute granules with characteristic staining properties, 1000–3000 Å in diameter, as shown in Fig. 14.2(a). This material can also be detected in the axons which pass back towards the corpora cardiaca, and if the axons are ligatured or severed, neurosecretory material accumulates above the point of interference (see Fig. 14.2(b)), suggesting that the granules are transported from cell body to corpus cardiacum by bulk flow along the axons. At the same time that neurosecretory material dams up at the level of ligatured axons, the store of such material which is normally present in the corpora cardiaca disappears, and it seems that this organ is mainly concerned with the storage and release of brain hormone.

The role of brain hormone in moulting was demonstrated by the early experiments of Kopeč, who showed that when larvae of the gypsy moth were deprived of their brains 10 days or more after the last larval moult, pupation

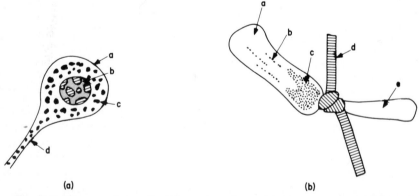

(a)   (b)

Fig. 14.2. Neurosecretory cells of the protocerebrum. (a) A neurosecretory cell from the locust. a, cell body; b, nucleus; c, neurosecretory granules; d, axon (redrawn from Highnam, 1961). (b) Schematized drawing of fresh preparation of the fused nervi corpora cardiaca and nervus recurrens of the blowfly three days after ligature. a, proximal part of nerve; b, axon with swellings of neurosecretory material; c, bulk accumulation of neurosecretory material; d, ligature; e, distal part of nerve, lacking neurosecretory material (drawn from photomicrograph of Thomsen, 1954).

occurred, and brainless but otherwise normal pupae and adults were formed. If the extirpation was carried out less than 10 days after the moult, however, the caterpillars failed to pupate, although they continued alive for a long time. Larvae which had been tightly ligatured behind the thorax pupated if the ligature was tied after the 10th day, but before that only the front half underwent pupation. These experiments showed clearly that a blood-borne factor was involved in the induction of the moult, and that there appeared to be a critical period at about the 10th day, during which the secretion was liberated.

Later experiments have amply confirmed these original findings, and have served to implicate the neurosecretory cells of the protocerebrum specifically in this action. If this region of the brain, for instance, is implanted into insects which have been decapitated before the critical period, the implants will activate the system, and moulting will take place in insects which would otherwise not have moulted.

The chemical nature of the neurosecretory hormone has not yet been unequivocally established, but it seems likely that it is a peptide, since it can be inactivated by bacterial proteinase. The stimuli which cause its liberation appear to vary greatly between different species of insect; in some they are closely linked with the act of feeding, which is not surprising in view of the heavy demands on raw materials which are made by processes of growth; in these, the effect seems to be mediated, at least in part, by stretch receptors associated with the alimentary canal, monitoring the distension caused by the presence of food. But in other species other factors appear to be involved, and the situation does not lend itself to convincing generalization.

b. The Thoracic Glands

In the course of time it became apparent that it was not just the liberation of brain hormone from the corpora cardiaca that caused moulting. If certain species of insect were ligatured, for instance, in such a way that the head was isolated from the thorax before the critical period, neither the anterior nor the posterior portions moulted. It was subsequently discovered that a two-stage process is in fact involved, with the brain hormone exerting its effect on the thoracic gland, and with this in turn producing and liberating a moulting hormone.

The thoracic glands originate as ectodermal invaginations of the head region, but they take up different positions and assume different forms in different insects. In some they remain in the head as compact "ventral glands"; in some they become associated with the corpora cardiaca/corpus allatum complex as a "Weisman's ring"; in some they form a loose network of cells in the thorax, while in many they become closely associated with the tracheal system. Whatever their position, when cells of the thoracic gland come under the influence of the brain hormone, they enter upon a cycle of activity involving an increase in the size of the nuclei, which often become extensively lobulated, and an increase in the amount of basophilic material in the cytoplasm (see Fig. 14.3(a)). The precise significance of these changes has not yet been established, but it appears that the cells become involved in the synthesis of the moulting hormone itself, together with a variety of mucoproteins, glycoproteins and other substances with which the hormone appears to be associated. The hormone itself has been isolated in crystalline form, with a yield of about 5 mg from 100 kg of

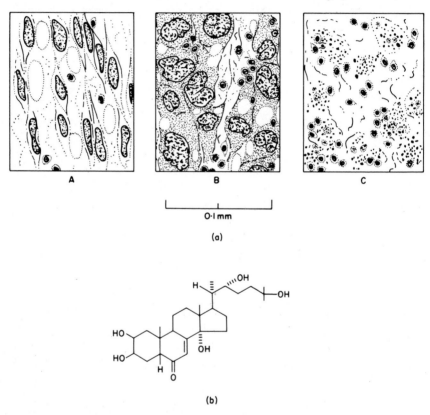

Fig. 14.3. The thoracic gland. (a) Changes in the histology of the thoracic gland during its cycle of activity in *Rhodnius*. A, the inactive stage in unfed fifth-stage nymphs; B, the active stage in fifth-stage nymphs 10 days after feeding; C, the phase of regression in adults one day after moulting, showing numerous haemocytes around the disintegrating nuclei (Wigglesworth, 1952). (b) The structure of the moulting hormone α-ecdysone (from Karlson and Sekeris, 1966). (c) The relationship between the quantity of moulting hormone injected into last instar larvae of *Chironomus tentans* and the degree of "puffing" of locus I-18-C on the salivary gland chromosome 3 hr after injection. The sketches show the appearance of the four classes of "puffing" distinguished, whose frequency distribution among larvae injected with the stated quantities of hormone are illustrated in the histograms below. The shift towards higher levels of "puffing" with increasing dose is clearly shown, n = number of larvae used (Clever, 1963).

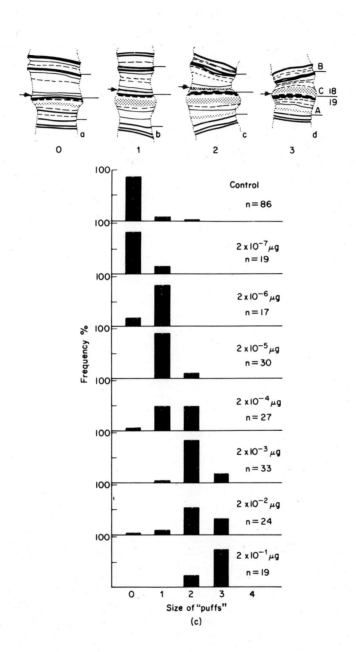

(c)

silkworm pupae. It has recently been identified as 2,3,14,22,25-pentahydroxy-Δ7-cholestene-6-one, and its structure is illustrated in Fig. 14.3(b). 7.5 mμg of the purified material is capable of inducing puparium formation in the isolated abdomen of blowfly larvae. Following production of the moulting hormone and associated materials, the cells of the thoracic gland enter upon a phase of regression, and revert to their resting condition.

The primary effect of the release of moulting hormone from the thoracic gland is an "activation" of all cells which are involved in the processes of growth and moulting, notably the cells of the epidermis. There is an enlargement of the nucleoli, a mobilization of ribonucleoprotein and an increase in the rate of protein synthesis. There is also an increase in the number of mitochondria, which must in itself involve a substantial synthesis of their protein and lipid components. The mechanism by which this activation of the synthetic machinery is achieved is still uncertain, but recent observations suggest that a direct action on the genetic material may be involved. It has been shown that within 10 min of the injection of moulting hormone, in quantities corresponding to about 10 molecules per haploid set of chromosomes, it is possible to distinguish the appearance of "puffs" on salivary gland chromosomes, an effect which may reflect the activation of particular genes (see Fig. 14.3(c)). The result of such activation might be the appearance in the cells of enzymes which could cause a shift in the metabolic pattern, such as to promote the formation of components required for the processes of growth and moulting.

c. The Corpus Allatum

The corpus allatum is a compact tissue composed of dense clusters of glandular cells; in the active condition, the cells are relatively rich in cytoplasm, and contain aggregations of glycoprotein granules. During the resting phase there is a decrease in cytoplasmic volume (as illustrated in Fig. 14.4) and the cell membranes tend to become deeply folded.

The corpus allatum is active in the control of reproduction (see below) and one of the most fruitful sources of the hormone that it secretes has proved to be the abdomens of adult, male giant silkworm moths of the genus *Hyalophora*. Extraction of abdomens with lipid solvents yields an oily, orange liquid, which shows high activity in commonly employed assay systems for juvenile hormone. If, for example, it is applied to punctures in the pupal cuticle of the mealworm beetle, it induces the formation of a larval type cuticle, instead of the adult type, at the next moult. The development of convenient assay techniques like this has facilitated the separation and purification of active materials from a variety of sources, including not only insects but vertebrate tissues, as well as bacteria and plant material. Activity was found to be associated with open-chain terpene components, and the active principle was originally identified as farnesol and its aldehyde derivative, farnesal; recent work, however, has established the identity of the active principle in *Hyalophora* extracts as the related compound, methyl *trans, trans, cis*-10-epoxy-7-ethyl-3,11,dimethyl-2,6-tridecadienoate, the structure of which is illustrated in Fig. 14.4(c) (Röller *et al.*, 1967). In its ability to ensure the retention of juvenile characters in a variety of assay systems this compound is more than a thousand times as effective as farnesol, and there seems no reason to doubt that it constitutes the true juvenile hormone.

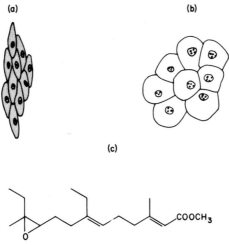

Fig. 14.4. The corpus allatum. a, histological appearance of cells of the inactive corpus allatum of *Leucophaea madera*; the spindle-shaped cells are closely packed and have little cytoplasm; b, histological appearance of cells in the active corpus allatum, showing a substantial increase in cytoplasmic components (drawn from photomicrographs of Engelmann, 1957/8); c, the chemical structure of the juvenile hormone (Röller *et al.*, 1967).

The effect of juvenile hormone in the developing insect is to ensure the production of larval, or juvenile, characters. This has been convincingly demonstrated by experiments in which the source of the hormone in young larvae of the silkworm is removed by operative procedures. In the absence of the corpus allatum, and hence of juvenile hormone, the larvae moult to form precocious pupae of minute dimensions, weighing no more than 2.5 mg, as compared with the normal weight of 1.25 g. The precise mode of action of the hormone has not yet been determined, but it may involve a direct effect on the gene system, such as to activate those elements of the total system that are associated with the formation of larval characters, at the same time suppressing the activity of those that are responsible for the formation of pupal and adult features. The effect seems to be graded, in the sense that the final outcome depends on the precise titre of juvenile hormone during the active period, as illustrated in Fig. 14.5. If the titre is high throughout the active phase, the immature characteristics are fully realized. If the titre is lower, or if high titres are not present until comparatively late in the cycle of activity, then a condition intermediate between the immature and the adult form may be produced; while if the titre of juvenile hormone remains low throughout the active phase of growth, then larval characters are completely suppressed, and those parts of the gene system that are responsible for the formation of adult characters are allowed full expression.

One very important aspect of the activity of the juvenile hormone is the maintenance of the structural integrity of the thoracic gland. This gland may, in fact, be considered as an immature character, since it is the moulting hormone which it produces that ensures the continuation of growth and moulting, an essentially juvenile process; without a supply of moulting hormone, this process would inevitably come to a stop. In the presence of juvenile hormone, the thoracic gland will, at the end of its cycle of activity, enter upon a resting phase in preparation for the next cycle. But if the thoracic gland goes through a cycle of activity in the absence of juvenile hormone, then the gland breaks down completely. Within 24 hr of the final moult in *Rhodnius,* for instance, the nuclei of the thoracic gland can be seen to be undergoing chromatolysis, and after 48 hr they have disappeared completely; the disintegrating fabric of the gland is concurrently invaded by phagocytic haemocytes, which help to dispose of the cellular debris (see Fig. 14.3(a)). Here it is not just a question of allowing the expression of one potentiality, while suppressing another, but rather of a complete shut-down of the gene system. With the disappearance of the thoracic gland, and the consequent cessation of growth and moulting, the corpus allatum is free to resume activity, and it now becomes effective in the control of reproduction rather than of development, as will be described below.

The mechanism by which the secretion of juvenile hormone by the corpus allatum is controlled has not yet been fully elucidated. Its activity appears to be low just before the larval moult, rising to a peak just after the moult; it becomes completely inactive after the pupal moult, and remains so until two-thirds of the way through the pupal period in the silkworm. The indications are that a restraining influence on the corpus allatum is exerted by centres in the brain, but a great deal more work remains to be done on this aspect of the problem.

On the basis of the results reviewed in this chapter, it would seem that the growth and development of insects could be interpreted, as a first approximation, in terms of an interaction between the system's genetic basis and two hormones, the moulting hormone and the juvenile hormone. The secretion of moulting hormone causes an activation of synthetic systems leading to the formation of new substance, as the indispensable basis for the process of growth; while the intensity and timing of secretions of juvenile hormone govern the type of substance that is formed, whether larval or adult (see Fig. 14.5). There is evidence, however, that other hormones may be involved in the control of certain aspects of the process. Hardening and darkening of the newly formed

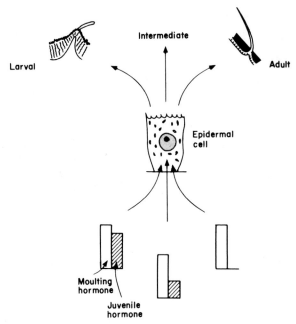

Fig. 14.5. Diagram illustrating the hormonal control of metamorphosis in epidermal cells of a hemimetabolous insect acted upon by moulting hormone (open rectangles) and juvenile hormone (hatched rectangles) in different proportions (redrawn from Wigglesworth, 1965).

cuticle, for instance, appear to be controlled independently of other processes involved in ecdysis. The active factor is released from neurosecretory cells of the brain, apparently in response to nervous stimuli from the thorax.

2. Neuroendocrine Control of Reproduction

The resumption of activity by the corpus allatum which follows the completion of development, and to which reference has already been made, appears to be an essential feature in the control of reproduction. In allatectomized females, oöcyte development only proceeds to the point at which yolk would normally be deposited, after which a process of resorption supervenes. The secretions of the corpus allatum appear also to play an important part in the male, with particular reference to the activity of accessory glands and the formation of spermatophores.

There are indications that the material secreted by the corpus allatum may serve a more substantive function than that of chemical messenger. This is suggested, for instance, by the close correlation which has been established between ovarian activity and corpus allatum volume, as illustrated in Fig.

14.6(a), which indicates that the gland might actually be furnishing raw material for the formation of yolk; and also by the fact that removal of ovaries has been shown to cause hypertrophy of the corpus allatum in several species, as if material which would normally be used for oöcyte development has under these circumstances had to be retained by the endocrine gland. It would be in accord, too, with the observation that male moths serve as a richer source of juvenile hormone than females, which contain only a fraction of the male activity; and

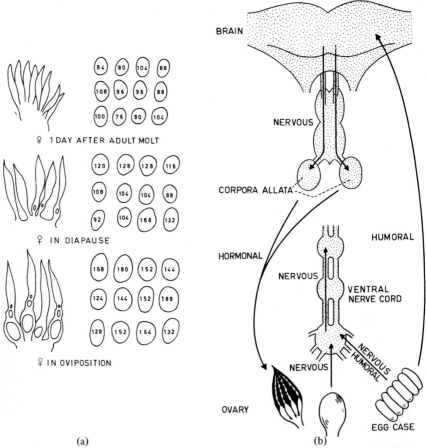

(a)   (b)

Fig. 14.6. Neuroendocrine control of reproduction in female insects. (a) The relation between ovarian activity and the volume of the corpus allatum in *Leptinotarsa* (de Wilde, 1964). (b) Diagram illustrating the factors involved in the control of reproduction in *Leucophaea*. Mating impulses are transmitted via the ventral nerve cord to the proto-cerebrum, resulting in a release of corpus allatum activity and subsequent induction of follicular activity. During pregnancy, the corpora allata are inhibited, presumably by nervous and humoral mechanisms resulting from the presence of an egg case in the brood sac (de Wilde, 1964 after Engelmann).

with the fact that implantation of ovaries into male pupae results in a marked decrease in the content of juvenile hormone, suggesting that the developing ovaries in some sense "consume" the active principle, whatever the precise significance of the utilization may be.

In many insects, the development of ovaries is closely linked with the nutritional state, as one might expect in view of the demands that oöcyte development would make on food reserves; in the absence of nourishment, oöcyte development may, in fact, be suppressed completely. In this effect, too, the corpus allatum has been firmly implicated, since the activity of the gland itself can be shown to be governed by nutritional state; and if active corpora allata are implanted into starving insects, ovary development commences.

While the corpus allatum appears to exert a major influence on oöcyte development, it is by no means the only important factor. Allatectomized females of *Calliphora* will occasionally produce fully developed eggs, but they fail to do so if the neurosecretory cells of the brain are removed. Reimplantation of corpora allata fails to restore ovarian activity in such insects, but implantation of neurosecretory cells, or of corpora cardiaca, leads to development of the ovaries. It would seem that the normal function of the corpus allatum is dependent on connection with the neurosecretory cells, and that in the over-all regulation of reproduction, there is a complex interplay of nervous and humoral factors, involving protocerebrum and lower nerve centres as well as corpus allatum, corpora cardiaca and the ovaries themselves. By virtue of such central nervous participation, ovarian activity will come under the influence of a variety of environmental factors associated with reproduction, and specifically related to the biology of the insect concerned. For this reason, it is difficult to generalize the situation, and it will be necessary to consider a concrete example in order to illustrate the interaction of factors involved in the control of reproduction.

One of the most thoroughly investigated species is a viviparous cockroach, *Leucophaea madera,* whose reproduction shows a particularly striking interaction of nervous and humoral influences (see Fig. 14.6(b)). In this species, the activity of the corpus allatum appears to come under the restraining influence of higher nerve centres. The act of mating releases the corpus allatum from nervous inhibition, an effect which is mediated through the ventral nerve cord, and the development of oöcytes proceeds to completion under the influence of its secretions. The disinhibited corpus allatum enters upon a phase of cyclical activity correlated with successive waves of oögenesis, suggesting the existence of some sort of feedback from the ovaries. Ovarian development is arrested during the period of embryogenesis, as the result of what appears to be activity in a dual control system; the presence of eggs in the brood sac seems to cause a nervous input to the ventral cord, to produce an inhibitory effect on the corpus allatum via the nervi corpora cardiaca; additionally, there appears to be a direct endocrine effect, because the disinhibition which can be produced by removing the egg case from the brood sac can be reversed by implanting the egg case in the body cavity; and severing of the ventral nerve cord is not as effective in activating the corpus allatum as is egg case removal. Following completion of embryonic development, the act of parturition provides an input to the ventral nerve cord which has the same effect as mating, and causes the corpus allatum to be released from inhibition, thus re-starting the cycle of reproductive activity.

3. Other Neuroendocrine Control Mechanisms

The regulation of growth and reproduction provides particularly striking examples of the working of neuroendocrine control mechanisms, but there can be little doubt that many other physiological activities are regulated, though perhaps less spectacularly, by similar means. It seems, indeed, that reproductive hormones may have quite general effects on the metabolism of adult insects, causing, for instance, an increase in the rate of oxygen consumption which has been shown to be independent of ovarian development. The question therefore has been raised, whether the control of egg production may not be simply one facet of a general effect on respiratory and synthetic metabolism. Certainly, the neurosecretory cells have been firmly implicated in the control of protein secretion by cells of the midgut in certain flies, and similar effects have been described for other kinds of insect. Their secretions appear also to exercise an important general influence on the protein metabolism of the locust, causing an increase in the level of blood protein by activating the synthetic machinery of the fat body. If these blood proteins fail to be taken up by the ovaries, there is a progressive rise in blood protein level. To what extent these, and similar, effects reflect a direct action of the hormones, as opposed to a general homeostatic adjustment between different components of the total somatic system, is uncertain at this stage, but there can be little doubt that they will prove a fruitful field for experimental investigation by modern biochemical techniques, and that the situation is capable of elucidation on this basis.

The problem of neuroendocrine control mechanisms has assumed a new dimension with the recent discovery, by Finlayson and Osborne, of the wide

Fig. 14.7. Peripheral neurosecretory cells of the stick insect. (a) Diagram of the innervation of one side of an abdominal segment. Neurones on the course of nerves are numbered from 1 to 10; fb, fat body; fbn, fat body neurone; mn, median nerve; na_1, na_2, na_{2a}, nl, nl_2, np_2, np, npa, nerves; nh, neurohaemal tissue; sp, spiracle; tns, transverse nerve swelling (Finlayson and Osborne, 1968). (b) Schematized diagram of a transverse section through a part of the link nerve (na_1 of (a)) in the stick insect. A group of nerve fibres, Nf, some of which contain granules, is contained by wrappings of the Schwann cell (Sc). A large nerve process (Np) is packed with granules and has no Schwann cell sheath. This "naked" process, whose membrane is markedly convoluted, is separated from the haemolymph (H) by only a thin coating of basement membrane material (Bm). N, Schwann cell nucleus. (Drawn from electron micrograph of Finlayson and Osborne, 1968.)

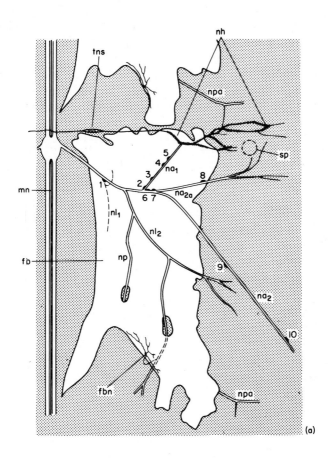

(a)

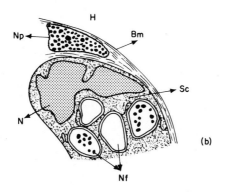

(b)

distribution of peripheral neurosecretory elements. At least 10 multi-terminal neurones have been identified on each side of each abdominal segment of the stick insect, as illustrated in Fig. 14.7(a), associated with the major nerves and tracheae. The neurone dendrites contain accumulations of typical neuro-secretory granules, and they are often separated from the haemocoele by no more than a thin coating of basement membrane material (Fig. 14.7(b)). There seems little reason to doubt that these granules are destined for release into the haemolymph, but further experimental work will be required to determine the factors which cause release, and the nature of the target of action. The picture which is suggested by these preliminary results, however, is of a diffuse pattern of neuroendocrine control, of which one demonstrated example would be the diuretic effect of hormones from neurosecretory elements associated with the abdominal nerve cord of *Rhodnius,* which has been mentioned earlier. It is possible that the diffuse nature and wide distribution of many neuroendocrine elements could usefully be seen as another reflection of the inefficient pattern of blood circulation which characterizes most insects (see Chapter 3). The corpora cardiaca could be interpreted as a condensed form of what is generally a diffuse system, situated in a region of maximum condensation of segments, and responsible for the regulation of activities whose time scale is so extended that efficient distribution of neurosecretory material would not be a primary requirement.

Aspects of Physiological Ecology

INTRODUCTION

The first three sections of this book have dealt with the physiology of insects as elucidated by experiments carried out under standardized and closely controlled conditions. What has been described, essentially, is the physiology of the individual insect in the laboratory; what could be considered to be of far greater interest and importance would be an account of the physiology of the species in its natural environment. Such an account would clearly have to be based on knowledge gained through an experimental analysis of different physiological processes under laboratory conditions, but it would go beyond this to a consideration of the way in which these processes interact in a complex and ever-changing environment; and in the final analysis, this interaction would find expression in population dynamics, with the shifting balance between birth and death determining the number of individuals present in the population at any one time. The influences which would have to be assessed in this context would include not only the physical factors of the environment, but also the interrelations between the insect and its parasites and predators, its symbionts and its competitors, whether intraspecific or interspecific. An account of population dynamics in such terms would perhaps represent a goal towards which all investigations of insect biology might be said to tend, but it is one which is quite unattainable on the basis of present knowledge. Even with the species that have been most thoroughly studied, both from the physiological and the ecological point of view, the complexities of the situation militate against anything more than a superficial evaluation of some of the more important factors. Nor could the concept of the "typical" insect be extended to a treatment of physiological ecology, because what is of importance here is the detail of the interaction between the insect and its environment, and this would depend so fundamentally on the particular species involved. When one considers, for instance, the diverse feeding habits of insects, ranging from the various general categories of omnivorous, herbivorous and carnivorous, through the more restricted types of sap-sucking, blood-sucking or nectar-feeding, to the highly specialized modes of feeding seen in the wax moth or the clothes moth, it is clear that the relations of insects to this particular aspect of the environment would be completely beyond meaningful generalization. The same would apply to most other features of the environment, particularly the biotic ones, but it seems possible to single out two which, because of their very direct effect on the physiology of insects, might be capable of treatment in general terms; the one, humidity, the other, temperature. Both are so closely linked to the basis of physiological function, that insects could be expected to be affected by them in a generally similar way, and these features have accordingly been isolated for discussion from the complex of factors which in their totality comprise the environment.

If the balance between the rate of birth and death of insects is affected by humidity and temperature, or more generally, by climate, then one would expect to see the seasonal and long-term changes which characterize terrestrial environments reflected in corresponding fluctuations in the population density of insects. That striking fluctuations do occur in the density of many species of insect has long been known, though data on this aspect of insect ecology have usually to be accepted with caution. The existence of serious sampling errors is recognized by all workers in this field, but it is not often that attempts have been made to evaluate their magnitude. However, even if the precision of population estimates is usually questionable, there can be little doubt that the kind of short-term and long-term changes that are illustrated in Fig. 15.1(a) do provide an indication, however imperfect, of the real changes in population density that underlie them. Such changes can often be plausibly, and sometimes even convincingly, related to corresponding changes in tememperature or humidity. It can be seen from the inset of Fig. 15.1(a), for instance, that the apparent density of tsetse flies tends to be low at the height of the hot-dry season, and significant correlations can be obtained between apparent density and temperature or humidity (or saturation deficit, which is effectively a combination of the two, see Chapter 16). Similar correlations between physical factors, like minimum temperature and rainfall, and the number of insects caught in light traps have been obtained in temperate climates, and Fig. 15.1(b) illustrates one of the more sophisticated examples of this approach to the problem of population dynamics. Here the relation between the population density of thrips and various environmental factors has been analysed by the statistical technique of multiple regression, and a close correspondence has been established between the observed population density, and the density which can be calculated on the basis of the regression equation, by substituting for recorded values of the physical factors. This demonstrates that, for the situation described, the density fluctuations can be almost completely accounted for by the climatic fluctuations. In the mind of the sceptic, however, the question inevitably arises whether the relation between the variables is a causal one; to return to the earlier example, one may ask whether the population density of tsetse flies is low *because* the saturation deficits are high in the hot-dry season, or simply *when* the saturation deficits are high. It could well be that both saturation deficit and density are correlated with some third factor, or complex of factors, and that it is with this that a causal relationship subsists. It is

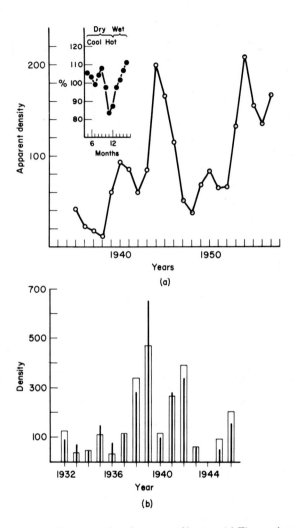

Fig. 15.1. Aspects of the population dynamics of insects. (a) Fluctuations in the density of tsetse flies. The main curve shows long-term changes in the apparent density of *Glossina swynnertoni* at Shinyanga, Tanganyika, over a period of 21 years. The inset shows seasonal variations about the annual mean, with monthly catches expressed as a percentage of that mean; a marked decrease in density occurs with the onset of the hot-dry season, followed by a rapid recovery during the rains. The apparent density represents the catch of adult males per 10,000 yards of transect, and bears a relation, though not necessarily a constant relation, to population density (redrawn from Glasgow and Welch, 1962). (b) A comparison between the observed density of thrips at the spring peak (open columns) and the density calculated on the basis of a multiple regression with climatic factors as independent variables (closed columns); for further explanation see text (Bursell, 1964b from Davidson and Andrewartha).

important to bear in mind that this doubt can never be answered by regression analysis; the regression coefficient can provide no more than an indication that a particular physical factor is important in relation to the population dynamics of a particular species, no matter how closely the situation is described by the regression equation. Unless that suggestion can be independently verified, it remains no more than a suggestion.

With this qualification, the available evidence does indicate that temperature and humidity may have a substantial influence on the population density of insects, and attention must be turned to the mechanisms by which such effects could be exerted. Any environmental factor which affects the rate at which insects are born to the population, or the rate at which they die, will have an effect on the numbers present at any one time, and hence on population dynamics. Consideration must accordingly be given, in the last two chapters of this book, to the ways in which the birth-rate and the death-rate of insects may be affected by temperature and by humidity.

CHAPTER 15

TEMPERATURE EFFECTS

In the introduction to this book it has been suggested that an insect can be considered essentially as a special type of metabolic system, and this point of view is of obvious relevance to the present discussion, because the reaction rates of component parts of such a system should be related to temperature in a fairly simple way. An example is given in Fig. 15.2(a), which shows the effect of temperature on the rate of a chemical reaction catalysed by an enzyme. It can be seen that, over the lower range of temperatures, the reaction rates increase by a factor of about 2 (the precise value may vary between 1 and 4 depending on the enzyme). At higher temperatures the situation is complicated by the thermolability of most enzymes; here the curve tends first to flatten out, as the increase in reaction rate is partially offset by an increase in the rate of thermal inactivation; and then to fall towards zero, as the enzyme becomes more quickly inactivated. A competition between two rate processes with opposite effects is involved here, and the temperature at which the quantity of reaction product is maximal (the optimal temperature) will depend on the duration of the assay, being lower for long than for short durations.

If an insect can be considered as a metabolic system comprising a network of metabolic pathways, each mediated by a sequence of enzymes, then the effect of temperature on any particular life process might be expected to take a form similar to that illustrated in Fig. 15.2(a). The close resemblance between the four curves of Fig. 15.2 suggests that this may not be too naïve a point of view. Fig. 15.2(b) shows the effect of temperature on the rate of oxygen consumption of an insect, which may be considered as a process mediated by a complex enzyme system rather than by a single enzyme. Figure 15.2(c) shows the effect of temperature on the rate of development of an insect, which may be considered as the end-product of an immensely complicated network of pathways. Figure 15.2(d) shows the relation between temperature and what might be called the rate of living, a process which cannot be precisely defined, but which can be estimated as the reciprocal of longevity, in the same way that the rate of development can be estimated as the reciprocal of developmental duration; it would denote, in other words, how long the insect takes to live its

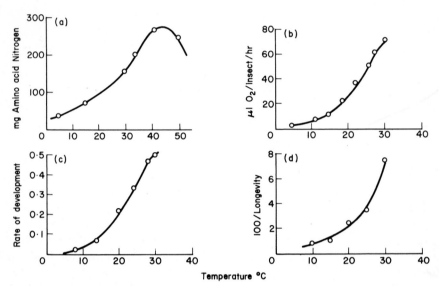

Fig. 15.2. The effect of temperature on life processes. (a) The effect of temperature on the activity of a digestive protease, expressed as the amount of amino acid nitrogen released by hydrolysis of a protein substrate (redrawn from Baldwin, 1948 after Berrill). (b) The effect of temperature on the rate of oxygen consumption of unfed tsetse flies, *Glossina morsitans* (redrawn from Rajagopal and Bursell, 1966). (c) The effect of temperature on the rate of pupal development in female *Glossina morsitans*, expressed as the proportion of total development completed in 10 days (drawn from data in Phelps and Burrows, 1969). (d) The effect of temperature on the longevity reciprocal of adult fruit flies, *Drosophila melanogaster*, as a measure of the rate of senescence (data from Loeb and Northrup, 1917).

life, how quickly it senesces. The general form of all the curves mirrors the simple enzyme relation of Fig. 15.2(a) quite accurately over the range that permits of their estimation. This range is limited by a discontinuity which is generally described as death, where the metabolic system ceases to exist as such. In Fig. 15.2(b) and (c), the curves do show a tendency to flatten out before this point is reached; in (d) no second inflection of the curve would be expected, since there would here be no question of opposing, but rather of reinforcing, rate processes; the thermal inactivation of proteins would contribute to, and probably take over from, other reactions as a direct cause of senescence and death, so that the rate of dying would tend to infinity rather than to zero.

The indication from these general relations are, therefore, that temperature would have a readily definable effect on the rates of metabolic processes leading to the birth and death of insects, and hence a predictable effect on the population dynamics of insects. Naturally, one would not suggest that all processes relevant to the birth and death of insects can be seen in such simple terms; nervous or humoral control mechanisms, which could be affected by temperature in quite a different way, would in many cases impose other patterns on the over-all relation with temperature. Such control mechanisms would, nevertheless, operate against the background of a general metabolic relationship with temperature, and what, therefore, would be of general importance would be the temperature of that metabolic system, of the insect itself, rather than of the environment; consideration must accordingly be given first to the factors which govern the temperature of insects.

a. The Temperature of Insects

In an insect which is in thermal equilibrium with its environment, and whose temperature is therefore constant or fluctuating slightly about a steady value, the net exchange of heat between the insect and the environment is zero, the gain of heat exactly balanced by the loss of heat. Losses would be occurring by long-wave radiation, by conduction and convection, and by evaporation; gains would be by solar and long-wave radiation and by metabolism. The metabolic component arises because part of the energy released during the oxidation of complex organic molecules fails to be captured in high energy phosphate linkage and appears instead as "metabolic heat" (see Fig. 1.1 of Chapter 1). The contributions made by these various processes to the heat flux at any given moment depend so much on circumstances, that it will be convenient to consider the problem in relation to a number of different situations.

With insects at rest in the absence of solar radiation, the body temperature is usually close to ambient, and for this reason neither long-wave radiation nor conduction and convection (which occur in proportion to the temperature differential) can play much part in the heat flux. Metabolic heat is the sole source of heat, and at the low rates which characterize the respiration of resting insects, it generally fails to raise the insect's body temperature by more than a fraction of a degree above ambient. The metabolic gains of heat can be effectively balanced only by evaporative cooling, and the precise point of equilibrium will therefore depend on factors which affect transpiration, being perhaps a fraction of a degree below ambient under conditions that favour evaporation, a fraction of a degree above in humid atmospheres. The differences are so small, however, that for all practical purposes, the insect may be considered to be at the temperature of the environment.

An exception to this generalization should perhaps be made where ambient temperatures are approaching the upper critical limit. There is evidence that some insects may then increase the rate of evaporative cooling by opening the spiracles (see Fig. 15.3(a)), and in this way ensure a lowering of body temperature to a few degrees below ambient. It is clear, however, that for animals as small as insects, water reserves would be inadequate to sustain long-term regulation of body temperature by evaporative cooling, and the process described should probably be seen as something of a crisis mechanism,

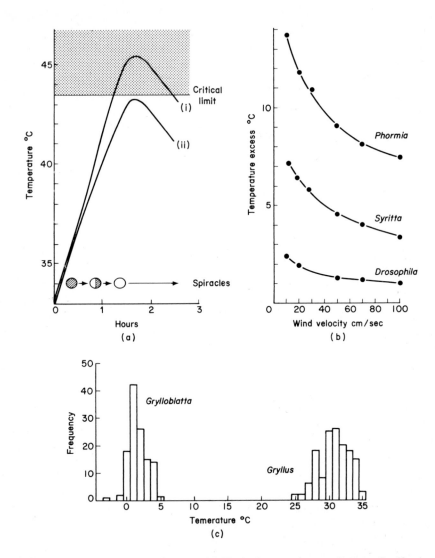

Fig. 15.3. The temperature of insects. (a) The body temperature of a tsetse fly *Glossina morsitans* exposed to ambient temperatures rising from 33° to 45° and then falling to 43°. Curve (i) shows the temperature during exposure in a saturated atmosphere; under these conditions, the insect's temperature differs by no more than 0.1-0.2° from ambient. Curve (ii) shows the situation when the insect is exposed in dry air, and it can be seen that up to a temperature of 35° there is little difference between the two curves, but at higher temperatures curve (ii) drops to a position about 2.0° below curve (i), indicating that evaporative cooling becomes a factor under these circumstances. Simultaneous observation of the spiracles showed that they were held closed until the temperature reached about 35° (hatched circles at bottom of graph), with "fluttering" (half-hatched circles) at higher

brought into play to stave off, for a limited period, the point at which tissue temperatures reach the critical limit.

When insects at rest are exposed to sunlight, the heat input from solar radiation becomes a dominant element in heat balance, and body temperatures may rise well above ambient. As substantial differences in temperature develop between the insect and its surroundings, convection becomes of importance in determining the equilibrium value, losses by conduction, evaporation and long-wave radiation playing relatively little part. The factors of importance in determining the equilibrium temperature will, therefore, be those that affect the rate at which heat is gained by radiation and lost by convection, and the most important of these are:

(i) size (see Fig. 15.3(b)); the larger the insect the greater the temperature excess to which it attains at a given level of radiation input. The insect's shape, and its orientation in relation to the sun's rays are also of considerable importance, while colour appears to be relatively unimportant;

(ii) the occurrence of air movements, which determine whether heat loss occurs by natural convection, relatively inefficient as a mechanism of cooling, or by forced convection, which is much more effective (see Fig. 15.3(b)). In this context, the nature of surface covering is also of considerable importance; a dense coat of hair or of scales, as in certain moths for instance, will greatly reduce the loss of heat by forced convection.

With insects in flight, a special situation arises, because of the enormous increase in metabolic rate, and hence in the rate at which metabolic heat is generated; temperatures of the thorax may increase to as much as $12°$ above ambient in flying insects. Despite the opening of spiracles and the vigorous ventilation of the tracheal system that occurs during flight, evaporation appears to play little part in dissipating the metabolic heat, most of which is lost by forced convection from thoracic surfaces. Since flight is a very intermittent activity in most insects, it seems likely that the changes in body temperature with which it is associated may be too ephemeral to affect mean tissue temperatures substantially, and they would therefore be of little importance in relation to the rates of birth and death in insect populations.

This brief review of the factors that affect the temperature of insects suggest that they have few powers of direct regulation of body temperature, which is

---

temperatures giving way to sustained opening (open circles) above $40°$. Exposure of tsetse flies to temperatures in excess of $43°$ for 1-2 hr usually causes death, and the graph shows that by evaporative cooling the insect is capable of maintaining its body temperature below this critical value, under the conditions shown (schematized from Edney and Barrass, 1962). (b) The effect of wind velocity on the temperature excess of large *(Phormia)*, medium *(Syritta)*, and small *(Drosophila)* flies, exposed to a radiation of 1.5 cal/cm^2/min (Bursell, 1964 from Digby). (c) The distribution of *Grylloblatta* and *Gryllus* in a gradient of temperature at high relative humidity (Bursell, 1964 after Jakovlev and Krüger, and Henson).

perhaps not surprising in view of their small size. The only source of heat which could be considered to be under the insect's direct control is metabolic heat, and because of its small size, the surface area from which heat is dissipated by convection is large in relation to the mass of respiring tissue, so that, except under the special circumstances of flight, the temperature excess which can be maintained is slight. Similarly, the only way in which insects could directly regulate losses of heat would be by increasing evaporation, and their scant water reserves would militate against the sustained use of such a mechanism. But the fact that they have minimal powers of direct regulation does not imply that they are in any sense at the complete mercy of the environment, for they possess considerable powers of indirect regulation. Even in the absence of solar radiation, most terrestrial environments are extremely heterogeneous in respect of temperature, and there may be substantial differences between closely adjacent parts. It has been mentioned that insects possess sense organs which are capable of responding to changes in temperature, and there is abundant evidence that such sense organs are put to use in selecting for habitation those parts of the general environment which are most suitable from the point of view of heat balance. Examples of the quite narrow thermal preference shown by certain species of insect are illustrated in Fig. 15.3(c); one of the species tends to aggregate in cool parts of a gradient, the other in warm parts. In addition to being able thus to take advantage of such temperature differences as may exist in the environment, many insects habitually bask in the rays of the sun, and by this means maintain a body temperature well in excess of ambient. These considerations suggest that the mean temperature of an insect's tissues may be widely different from, and probably far more favourable than, that of the general environment. It is clear, too, that to make even an approximate assessment of the mean temperature experienced by individuals of a population in a given species would pose a formidable technical problem. Yet it is on such an assessment that one would have to rely in any attempt to gauge the effect of temperature on population dynamics. Until such time as suitable methods are devised for the continuous monitoring of tissue temperatures in representative samples of a population, it will be impossible to make more than a general guess at the part that temperature may play in the regulation of insect numbers. The basis of even such a general guess would have to be a knowledge of the effects of temperature on various physiological processes, and it will therefore be useful to discuss the general way in which such effects would operate.

b. The Effect of Temperature on Birth-rate

The two main ways in which temperature influences the birth-rate of insects is through effects on the rate of reproduction and on the rate of development. Both appear to be rather delicate processes, to judge by the fact that they occur over a rather narrower range than many other physiological processes. The extent of the range, and its position, varies considerably from species to species, but in all there is a tendency, as the temperature increases from the lowest level which will permit the processes to occur, for the rate of oviposition and speed of development to increase towards a peak near the upper extreme. The rate of oviposition usually declines well before the upper limit is reached (see Fig. 15.4(a)), falling gradually to zero. The rate of development, on the other hand, seldom has a clear optimum, but tends to level off, or decline slightly, just before the upper extreme is reached (see Fig. 15.2(c)).

In their influence on birth-rate, these two effects will reinforce each other over lower parts of the range in the sense that, if an increase in temperature from 15° to 25° leads to an increase in the rate of oviposition from 0.2 to 1.0 eggs per day, and to a decrease in the duration of development from 20 days to 4 days (associated with a corresponding increase in the speed of development), then the rate at which adults are added to the population is effectively increased from $0.2/20 = 0.01$ to $1.0/4 = 0.25$ individuals per day. In other words, a five-fold increase in both rates leads to a 25-fold increase in the effective birth-rate. It is, of course, not legitimate to use such a simple approach, except to give a general indication of the magnitude of effect which may be expected, since it ignores the possible effect of temperature on associated factors, such as the length of the reproductive period or the mortality of developmental stages. The example simply serves to indicate that the rate at which adults will be recruited to a population of insects will be very markedly affected by temperature, and the relation with temperature would be of the general form described.

c. The Effect of Temperature on Death-rate

In the context of death it is useful to consider two types of effect; those which operate at extremes of the range, to kill the insect directly by processes which are as yet imperfectly understood; and those which operate within the viable range of temperatures, either directly, as in the phenomenon of senescence, or indirectly, in ways which will be described below.

(i) Critical Limits of Temperature. During the last few decades, advantage has been taken of the development of suitable statistical and experimental techniques to make a careful assessment of the critical limits of temperature in various species of insect. Much of the earlier work in this field suffered from a failure to evaluate in detail the relation with time, a relation which must be considered of paramount importance. At high temperatures a process, possibly of protein denaturation, is involved, which has an extremely high temperature coefficient, with correspondingly marked changes in time course over a narrow range of temperatures near the critical limit; while at low temperatures, where the formation of ice crystals may be a cause of death, the probability of occurrence of favourable molecular configurations will be a close function of

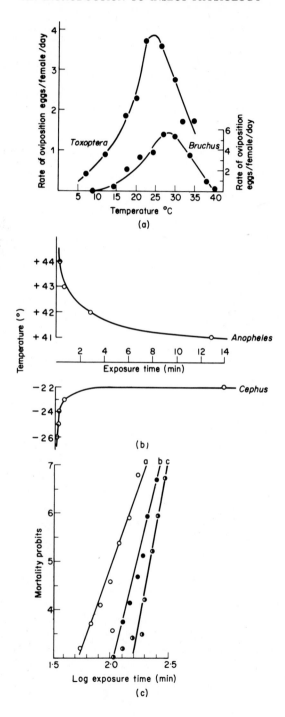

(a)

(b)

(c)

time. To get meaningful results, it is therefore necessary to study the thermal death point with minute reference to the duration of exposure. This is illustrated in Fig. 15.4(b), which shows the relation between temperature and the duration of survival at upper and lower limits for two different species of insect. Had an exposure of, say, 10 min been adopted as standard, the data show that the critical temperature for the mosquito would have been assessed as 41°, while for an exposure of 1 min, the value would be 2° higher, and similarly for *Cephus* at the lower range. In view of this relation, the best basis for evaluation would probably be to expose the insects to a given temperature near the critical limit for a range of durations, which would give a series of "dosage" levels at each of which the percentage mortality is assessed. The results can then be conveniently analysed by statistical techniques which have been developed in connection with toxicity tests, to give an accurate estimate of the time required to ensure the death of 50% of the individuals. An example of results obtained in this way is shown in Fig. 15.4(c).

The phenomenon of acclimation is one which must be taken into account in any study of critical temperatures. It has been shown for several species of insect that the critical temperature may vary according to the thermal history of the population from which the sample is taken. This is illustrated in Fig. 15.4(c), which shows that insects reared at 23° show 50% mortality (the corresponding probit value is 5.0) when the exposure time to a test temperature of 43° is 105 min, while half the individuals reared at 29° can survive for 154 min under these conditions; it shows also that a brief conditioning exposure to a temperature of 36° on the day before testing is even more effective in raising the resistance than is prolonged maintenance at 29°.

The carefully standardized work which has been done during recent years has lent precision to earlier indications that, for relatively short exposures of up to an hour or so, the upper critical temperature for insects in general is in the region of 40-45°, with tropical species usually more resistant than temperate species, and with certain specialized forms capable of withstanding extraordinarily high temperatures, in excess of 50°.

Over the lower range, where result may also be affected by the phenomenon of acclimation, the situation is complicated by the existence of different types of insect, differing in the nature of their response to low temperature. Some are

---

Fig. 15.4. Effects of temperature on the rates of birth and death. (a) The rate of oviposition of *Toxoptera graminum* and *Brucus obtectus* as a function of temperature (Bursell, 1964 from Wadley and from Menusan). (b) The relation between duration of exposure and the temperature at which 50% of individuals die; upper curve, *Anopheles* exposed near the upper critical limit (redrawn from Bursell, 1964 after Platt, Collins and Wilberspoon); lower curve, *Cephus* exposed near the lower critical limit (redrawn from Bursell, 1964 after Salt). (c) Mortality curves for *Dahlbominus fuscipennis* exposed to a temperature of 43°C at high relative humidity; (a), reared at 23°; (b), reared at 29°; (c), reared at 23° and exposed for 2 hr to 36° on the day before testing (Bursell, 1964 after Baldwin).

killed by exposures of about an hour to temperatures well in excess of zero, the cause of death being as yet uncertain. Others can withstand sub-zero temperatures so long as the formation of ice crystals in the tissue fluids is prevented; disruption of the submicroscopic architecture of the cells, which would occur under these circumstances, would probably constitute the immediate cause of death; for such species, the lower critical limit would be set by the point to which their tissue fluids would be capable of super-cooling. The third category of insect is capable of surviving the formation of ice crystals in their tissues; this ability appears to be associated with the presence of high concentrations of glycerol, which may accumulate to a level of 25% of the total wet weight of the insect. The mechanism by which glycerol exerts its protective effect has not been unequivocally established, but there are indications that the lethal factor may be the marked increase in the concentration of salts which occurs as ice crystals form, rather than mechanical damage, and that glycerol serves to "buffer" the tissue fluids against this concentration effect (e.g. Lovelock, 1953).

By virtue of the existence of these three general types of cold-hardiness, the range of lower critical temperatures for the group as a whole is much higher than for the upper critical limits, extending from well below $-35°$ in several species of hibernating insect from temperate and arctic regions, to temperatures above $+5°$ for tropical species like the tsetse fly.

(ii) Effects Within the Viable Range. The criterion used for determination of upper and lower critical limits of temperature is usually death within a limited period of time following exposure. It is likely, however, that temperatures which are not critical in this sense may yet materially affect the subsequent expectation of life. If this is so, then the distinction between effects at the limits of the viable range and within the viable range may be to some extent arbitrary, reflecting no more than a habit of considering the problem in terms of mortality rather than of survival. It is certainly clear from the results shown in Fig. 15.2(d), that temperatures within the viable range may have a profound effect on the longevity of insects, indicating that what one sees at the extremes of the range may be, in part, extensions of effects which are exerted in smaller measure at other points of the range. It may not be too uncharitable to suggest that the tendency to concentrate on upper and lower critical limits may be an expression more of the technical ease of the experimental procedure than of the intrinsic importance of such limiting effects. It is much easier to determine how many insects have died after a short exposure to a test temperature than it is to trace the development of sub-lethal influences resulting from prolonged exposure to sub-lethal temperatures.

Apart from the direct effect of high temperature in curtailing the expectation of life, there may be indirect effects which could be of great importance under natural, as opposed to laboratory, conditions; these would be associated with the increase in metabolic rate which occurs when temperatures are raised, as illustrated in Fig. 15.1(b). The rate at which oxygen is consumed reflects the rate at which food reserves are expended, and under conditions where the replenishment of reserves is intermittent, and the availability of food uncertain, a marked effect on life expectancy would be likely. To take a concrete example of what may be a general condition, the average fat reserves of a tsetse fly in the field would sustain life for about 5 days at a temperature of $20°$, but for only 2 days at a temperature of $30°$. In the cold season, the time available to the insect for locating a host animal from which food reserves could be replenished would therefore be twice as long as in the hot season and, other things being equal, so would its chance of finding a suitable host before dying of starvation, and thus of extending its life span by another hunger cycle. Indirectly, therefore, the expectation of life might be a close function of temperature.

d. Conclusion

The results presented in this chapter have demonstrated that through its direct influence on various metabolic processes, and on the nervous and humoral control systems which regulate the processes of reproduction, temperature may exert a profound effect on the rate at which insects are born and die to their populations. It is possible to gain some insight into the integrated effect of all these influences under laboratory conditions by the construction of life tables for a laboratory colony, from which age-specific fecundity rates can be calculated. From such data, the innate capacity for reproduction (denoted by the symbol r_m) can be estimated over a range of temperatures; this measures the rate at which females of reproductive age are added to a population of stable age distribution. Thus, instead of getting estimates separately of fecundity (Fig. 15.4(a)), speed of development (Fig. 15.2(c)) and length of life (Fig. 15.2(d)), a measure is obtained which represents an integration of these, and other, relevant processes in terms of reproductive potential. In the few species where this has been done, the relation with temperature is roughly as one would expect from a consideration of the component processes, with an inflected rise in the value of r_m from low levels at low temperatures to a peak, followed by a steep fall towards upper limits of the range. Unfortunately, this elegant approach to the problem suffers from a number of severe limitations. In the first place, it can only be confidently employed when the insect concerned is fully amenable to laboratory maintenance, and this is not the case with most species. Secondly, the evaluation is made under conditions which differ widely from those that prevail in the natural environment, particularly in such respects as the availability of food and water. Where these constitute limiting factors to survival, the results of laboratory investigations cannot be applied with any confidence to the natural state; for they are obtained with populations which are allowed to live their full life span, while, under natural conditions, causes of death other than old age are likely to exert a dominating influence on life expectancy.

Even if an accurate assessment cannot be made of the precise effect of temperature on reproductive potential, there can be little doubt that the general relation between them will be of the type described. It is known that the temperature of terrestrial environments is subject to substantial diurnal and seasonal variation, and may swing in the course of weeks or months between upper and lower limits of the viable range. Through their behavioural reactions to temperature, and by virtue of the heterogeneity of terrestrial environments, insects would be able to inhabit those parts of the environment that are most suitable to their life processes; and where they cannot escape exposure to sub-optimal conditions, they can to some extent buffer themselves against the effects of such exposure by their powers of acclimation. But despite these abilities, there can be no doubt that the seasonal changes in environmental temperature will be reflected, though to a degree which cannot at present be precisely gauged, in corresponding changes in the temperature of the insects themselves, and hence that corresponding effects on their population dynamics are likely to result. A proper insight into the quantitative details of such effects must await substantial advances in the investigation of physiological as well as ecological aspects of the problem.

CHAPTER 16

HUMIDITY EFFECTS

The relation between environmental humidity and the population dynamics of insects differs from that which has been described between temperature and population dynamics, because with humidity there is no question of a viable range, or of critical extremes. Humidity has no direct effect on the metabolic system in the way that temperature has, and extremes of humidity do not, in themselves, kill an insect. The effect that humidity has on insects is mainly an indirect one, through water content; if that is reduced below certain critical limits by exposure to desiccating conditions, or, conceivably, if it is raised above a certain limit, then the insect dies; and short of these limits a variety of sub-lethal influences may be exerted on processes capable of affecting the rates of birth and death. What is of importance, therefore, is to see water balance as a cumulative function, and in this respect unlike heat balance, which is essentially an instantaneous function. The differential between water losses and water gains must be regarded as adding up over a period of time, with a gradual depletion of reserves, perhaps, until the critical limit is reached. And in so far as this is the case, the important feature of the environment is not the amount of water vapour present in the atmosphere, its percentage saturation as expressed by relative humidity, but rather the amount of water vapour which is lacking from the atmosphere; for it is this that will determine its power to promote the evaporative loss of water. Unfortunately, the evaporating power of an environment is not a characteristic that can be easily measured. In the first place, it is closely related to temperature, since the saturated vapour pressure of air increases greatly with an increase in temperature, and so, therefore, does the saturation deficit, as illustrated in Fig. 16.1. The saturation deficit is a measure of the difference between the amount of water vapour present in the atmosphere and the amount which would be required to saturate the atmosphere at any particular temperature, and to this extent it provides a useful indication of evaporating power. At any given saturation deficit, however, the evaporating power may be substantially influenced by air movements, which will tend to steepen gradients of water vapour between the evaporating surface and the general atmosphere. This effect will be especially important when rates of

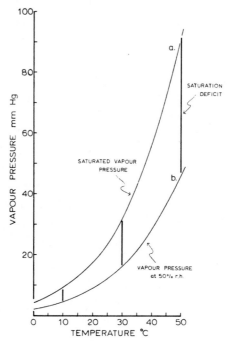

Fig. 16.1. The relation between vapour pressure and temperature at full and at half saturation. The saturation deficit at any temperature is given by the vertical distance between the two curves. (Bursell, 1964.)

evaporation are relatively high ("vapour limited" as opposed to "membrane limited" systems in the terminology of Beament, 1961). The general level of evaporation is of importance also in another respect, because it will govern the temperature of the transpiring surface, which at high levels of evaporation may be several degrees below ambient. It is clear that the evaporating power of a given environment, as it governs the rate of water loss from a given insect, cannot easily be assessed by measurements of the relevant physical features of the environment. Perhaps the best approach to the problem is the more direct one of measuring the rate of evaporation from a simple physical model, designed to simulate the organism under investigation in respect of surface area and general level of water loss. This method has been used with considerable success in an investigation of the spruce bud worm in Canada (Wellington, 1949), but most workers have been content to arrive at an approximation to the evaporating power by basing their results on measurements of saturation deficit.

In attempting to define the ways in which the evaporating power of the environment may affect the survival of insects, what needs to be considered first of all is, what exactly is the critical level of water content, or better, what is the extent of the water reserve, defined as the difference between the water content of the critically depleted and the fully hydrated insect, with allowance made for the production of metabolic water (see below). Secondly one would require to know what are the factors that influence the rate at which insects lose and gain water. It is in the relation between the extent of the reserve and the rate of its depletion that the effect of humidity on the rate of death must primarily be sought. In so far as humidity may affect the rate at which eggs are laid, and at which development proceeds, it will also have an effect on the rate of birth, but here the relation cannot usually be interpreted simply on the basis of a depletion of water reserves, and the problem must be tackled empirically.

1. The Water Reserves of Insects

The water reserve of an insect may be defined quite simply as the amount of water that the insect can afford to lose before the critical level of water content is reached. It should, on the face of it, be a simple matter to determine this quantity experimentally, but unfortunately the problem is complicated by a technical difficulty, so that, in fact, little accurate information is available in the literature on this point. Since insects are able to lose a certain proportion of the water that they contain, the actual amount of water that can be lost will obviously depend on their size. This applies both within species and between closely related species, as illustrated in Fig. 16.2. In order to determine the quantity of water reserve, allowance must therefore be made for differences in size between individuals in a sample, or for differences in mean size between samples. This is usually done by expressing water content as a proportion, or percentage, of some function of size, usually total fresh weight. Unfortunately, this procedure introduces a number of complications. It has been shown, for instance, that fat constitutes one of the most important food reserves in many insects; it would be possible to maintain an insect of this kind under conditions where losses of water were exactly counterbalanced by gains; the amount of water present in the insect would be constant, yet its water content, expressed as a percentage of fresh weight, would show a steady increase as the amount of fat decreases. Such anomalies could be partly circumvented by expressing water content as a percentage of the non-fatty dry weight, as has been done in Fig. 16.2, but this does little to improve the situation if substantial quantities of non-fatty reserves are also expended. The best way to overcome these difficulties would be to use some constant linear dimension as a measure of size, but this has seldom been done, and available data cannot, therefore, be used as a basis for an accurate assessment of the quantity of water reserve in insects generally. All that can be said is, that the water content of fully hydrated adult insects is usually in the region of 75% of the non-fatty dry weight, while critical levels of water content are in the region of 60-64%. This means that losses can be sustained of just about half of the water present in the fully hydrated insect (Fig. 16.2). The

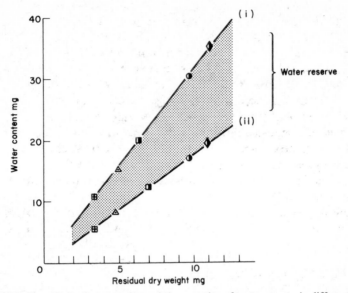

Fig. 16.2. The relation between size and the quantity of water reserve in different species of tsetse fly. Curve (i) shows the water content of flies which have newly emerged from pupae maintained at high relative humidity; curve (ii) shows the water content of flies that have been desiccated to the point where they are no longer capable of righting themselves. The difference between the two curves shows the extent of the water reserve, which is about 45% of the water originally present. ⊞, *Glossina austeni*; ▲, *G. palpalis*; ◨,*G. pallidipes*; ◑, *G. brevipalpis*;◆, *G. fuscipleuris*. (Simplified from Bursell, 1959.)

survival of the insect will depend on its ability to maintain a balance between losses and gains of water such that the reserve of water never becomes fully depleted.

2. The Water Balance of Insects

Insects can lose water by transpiration and by excretion, and they can gain water by ingestion and by the production of metabolic water. Some information relevant to a consideration of water losses and water gains by these processes has been given in earlier sections of this book (see Chapters 1, 5 and 7); in this chapter an attempt will be made to bring together the information on this subject, so as to provide a general picture of water balance in terms of the interaction of component processes.

a. Losses of Water

It is necessary to reiterate at the outset the difficulties that beset insects, in view of the fact that they are small animals inhabiting a terrestrial environment;

animals whose surface area, through which water vapour can be lost to the general environment, is large in relation to the water reserves which must sustain that loss. A primary requirement for animals faced with this difficulty would be an effective water-proofing of the surface, which would serve to reduce the loss by transpiration. A limit would be set to the extent to which an animal could thus insulate itself from the desiccating influence of the environment, by the need to provide for respiratory exchange, and for the excretion of waste products. These exchanges would have to be superimposed upon the general background of impermeability to water, and one would expect to find, associated with these points of exchange, the development of stringent control mechanisms to minimize the losses of water associated with them.

The permeability of the cuticle of insects to water is, in fact, extremely low, and water-proofing has been shown to be associated with the presence of lipids in the epicuticle (see Chapters 1 and 13). The rate of transpiration is, therefore, enormously increased if the epicuticular lipids are disrupted by exposure to lipid solvents or detergents, or to the action of abrasive dusts. It appears that the main barrier to the diffusion of water is provided by an orientated monolayer of lipid molecules, situated at the surface of the cuticulin layer, supposedly with their polar groups associated with the hydrophil surface, and their non-polar hydrocarbon chains extended outwards at an angle to that surface. It is this involvement of lipids as the basis of water-proofing that imposes a characteristic relation between temperature and permeability (see Fig. 16.3); rates of water loss show a sharp increase when the temperature of the transpiring surface reaches what appears to correspond to the melting point of cuticular lipids. This transition is presumably associated with an increase in the mobility of the lipid molecules, leading to changes in molecular spacing. It is unfortunate that it has not yet proved possible to relate this interpretation to the actual composition of the cuticular lipids. Cuticular extracts have always been found to comprise a complex mixture of lipids, only some of which conform to the requirements of the model for molecules with one end polar and the other non-polar. It is possible that it is only that fraction of the total which answers to this requirement that contributes to the orientated monolayer, but if that is so it is difficult to understand why there should be such a close correspondence between the transition temperature and the melting point of the total mixture.

The breakdown of water-proofing at the transition temperature is probably of limited biological significance, since in most insects it occurs at temperatures to which they would not normally be subjected, and in many it is above the upper critical limit. What is of interest from the point of view of general, as well as of molecular, biology are the enormous differences in permeability which characterize different species, as shown in Table 16.1. Usually the lowest values are found among insect eggs, where water reserves are extremely small in relation to surface area, or among pupae, where prolonged exposure to low humidities

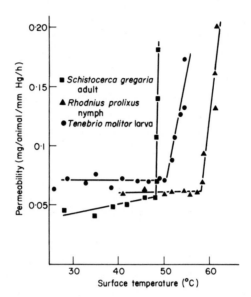

Fig. 16.3. The effect of temperature on the permeability of insect cuticles to water. The rate of transpiration per unit of saturation deficit remains relatively constant as temperature is raised until the "transition temperature", which differs from species to species, is reached. At this point there is a sharp and progressive increase in permeability as the temperature increases. (Beament, 1959.)

may occur under circumstances where a replenishment of water reserves is precluded. Active stages, whether larval or adult, tend to be comparatively poorly water-proofed. Within a given developmental stage, it is often possible to establish a correlation with habitat; those insects that inhabit arid environments, like the locust and the tsetse fly, being generally better water-proofed. Thus it would seem that differences between species and stages are in a broad sense adaptive, and what would be of interest would be to determine the nature of the adaptation at the molecular level, to determine what feature it is of the interaction between the lipids and their cuticular substrate that makes a difference of as much as three orders of magnitude in the cuticular permeability.

Because of the large surface area which characterizes insects generally, the losses of water sustained by cuticular transpiration constitute a substantial fraction of total loss, despite the efficient water-proofing of the surface. In a tsetse fly at rest in dry air, for instance, approximately 65% of the total water loss occurs by transpiration through the cuticle; the relatively low level of loss by transpiration from the tracheal system (20%) and by excretion (15%) is a reflection of the stringent control mechanisms which operate at these points, as already described (Chapters 5 and 7). During flight activity, the contribution from the tracheal system increases, and of the increased total loss only 30% can

TABLE 16.1

The permeability of insect cuticles, as determined by measuring the rate at which water is lost by cuticular transpiration in dry air at temperatures between 20° and 30°

Developmental stage	Genus	Permeability mg.cm^{-1}.hr^{-1}.mmHg^{-1} $\times 10^4$	Author
Egg	Rhodnius	30	1
	Lucilia	150	3
	Phyllopertha	600	5
Larva	Tenebrio	100	7
	Nematus	200	8
	Hepialus	1900	8
	Agriotes	6000	8
Pupa	Glossina	3	2
	Tenebrio	10	4
	Agriotes	228	8
Adult	Rhodnius	120	4
	Glossina	130	7
	Schistocerca	220	6
	Calliphora	390	7
	Periplaneta	550	7
	Bibio	760	8

1. Beament, 1949; 2. Bursell, 1958; 3. Davies, 1948; 4. Holdgate and Seal, 1956; 5. Laughlin, 1957; 6. Loveridge, 1968; 7. Mead Briggs, 1956; 8. Wigglesworth, 1945.

be attributed to cuticular transpiration; similar values have been reported for the locust. At high humidities, when spiracular and excretory control mechanisms are less stringently applied, the proportionate contribution of cuticular transpiration again decreases, this time in the context of a fall in total loss. These results relate to insects whose cuticles are relatively well water-proofed (see Table 16.1), and it is likely that in others the cuticular contribution may be even higher, but available information does not enable accurate estimates to be made.

To get some idea of the resistance of an insect to the desiccating influence of the terrestrial atmosphere, it is necessary to consider the losses sustained in relation to the extent of water reserves. In the unfed tsetse fly, at rest in dry air, the daily rate of loss amounts to about 25% of the water reserve, so that the insect should be able to survive for about four days without replenishing its water reserves, provided food reserves are adequate to sustain life for this period. Allowing for variations associated with activity, this estimate is in reasonable accord with values determined experimentally by the simple procedure of confining the insects in a dry atmosphere without access to food, and seeing how long they remain alive. Similar experiments have been performed with a number of insect species, and they enable estimates to be made of relative resistance to desiccation. Such estimates, however, are of limited relevance to the problem under discussion, in so far as they relate to insects that have been artificially deprived of the opportunity to replenish their water reserves. Under natural conditions, the losses sustained under the desiccating influence of the environment would be capable of being partially or wholly balanced by gains of water, and it is to this aspect of the problem that attention must now be turned.

b. Gains of Water

Water reserves may be replenished either by drinking or by the ingestion of food containing a certain proportion of water, which will range from 70-90% in blood-sucking and sap-sucking insects to as little as 10-15% in species that live as pests of stored grain. In order to assess the part played by ingestion in the water balance of insects living under natural conditions, one would need to estimate the frequency and the extent of replenishment, and for most species this would pose a formidable technical problem. The only insects for which quantitative information is available are species with very specialized feeding habits. It has been shown, for instance, that the sole source of water for certain pests of stored products is their almost dry food; and that under conditions of desiccation, these insects digest only a fraction of the ingested food, which appears to be eaten in part for the sake of the water that it contains. In obligatory blood-suckers, like the tsetse fly, the only source of water is the blood of vertebrates, and it has been shown that these insects retain a greater proportion of the water of their blood meal if their tissues have been dehydrated before feeding, or if the size of the blood meal is subnormal. In the case of species such as these, where the precise composition and availability of food can be estimated, it is possible to place these findings in the context of the species' biology. But with most other insects, the situation is complicated by uncertainties concerning the quantity and nature of the food supply in the normal environment. Results are available which indicate, for instance, that locusts can maintain water balance in dry air provided that there is a plentiful supply of food and that the water content of this food is in excess of about 35% (Loveridge, 1970); but to what extent this requirement is met in the normal environment during the dry season cannot be stated on the basis of present knowledge.

Another way in which water reserves are replenished is in the process of cellular respiration, where the hydrogen of the organic molecules that serve as substrate is transported through a system of hydrogen carriers to eventual combination with oxygen, leading to the formation of metabolic water (see Chapter 1). In assessing the contribution which this source of water may make to total water balance, it is necessary to take a number of factors into consideration. In the first place, the fact that an increase in the production of metabolic water will entail an increase in oxygen consumption, and hence in respiration, suggests the possibility that the gain in terms of metabolic water may be to some extent offset by an increase in water loss from the tracheal system. Secondly, it should be borne in mind that the gain which occurs in the course of oxidation of organic substrates cannot necessarily be counted as net gain. If food is ingested as carbohydrate, and if carbohydrate serves as the main substrate for oxidation, then it may fairly be considered that the oxidation of 1 g of carbohydrate will entail a net gain of 0.55 g of water. The oxidation of fat will produce 1.07 g of water for every gram of fat, but in this case the net gain may be very much less. The oxidation of fat produces a high yield of metabolic water because the fat contains a high proportion of hydrogen, but if, for instance, the fat reserve has been synthesized by the insect from carbohydrate or protein raw materials, then much of the hydrogen has had to be introduced into the fat molecule in the course of its synthesis, and such incorporation has been at the expense of oxidizable hydrogen, which may be considered as the equivalent of metabolic water. In this case, while the total yield of water may be 1.07 g per g of fat oxidized, the net yield would be no greater than if carbohydrate had been oxidized.

These considerations militate against any simple comparison between the three main classes of food substance. As a reserve, and neglecting the ultimate origin of the reserve, fat would appear to have the advantage of a higher calorific yield and a higher yield of metabolic water, and in a number of insects, particularly among members of the Lepidoptera and Orthoptera that indulge in prolonged migratory flights, fat constitutes a major proportion of stored foods; but further work will need to be done on all aspects of the synthesis and utilization of food reserves if the precise adaptive significance is to be unequivocally identified.

c. Summary

The results reviewed in this section have shown that, in order to survive, insects must maintain their water content within certain critical limits. It would seem that regulation of spiracular and excretory losses of water may play an important part in achieving this object, with the general level of cuticular permeability determining the range of environmental conditions over which such regulatory powers are likely to operate successfully. If the integument is very permeable, regulation may be effective over the humid end of the range, but exposure to dry conditions would probably prove inimical to long-term survival of the species. If the integument is well water-proofed, regulatory powers may sustain the species even in the driest parts of the range. In view of the limited longevity of most insects in very dry conditions, and of the fact that such conditions are a regular feature of many terrestrial environments, it seems likely that, through an effect on water balance, environmental humidity may exert a marked effect on the death-rate of insects. Unfortunately, it is not possible to go much beyond a general statement of this sort at present, because so little is known about the quantitative aspects of water replenishment for insects living under natural conditions. Without this information it is not possible to make an estimate of the probability that members of a species would exhaust their water reserves before they have a chance to replenish them, and it is on such an estimate that an assessment of the corresponding mortality would have to be based. In the present state of knowledge, it would perhaps be more fruitful to make a more empirical approach to the problem, by determining the state of water balance in insects sampled from their natural environment. If, in a proportion of individuals, the water reserve is near to the lower critical limit, it could reasonably be inferred that death by desiccation is an important factor in the population dynamics of the species. Information of this type is available for the tsetse fly, but in this particular species the indications are that, by virtue of well-developed regulatory powers and of a low level of cuticular permeability, desiccation is not an important cause of death. It would be of interest to extend this type of investigation to species that are less well-adapted to arid environments. Until this is done, assessment of the effect of humidity on death-rate will have to rely on the indirect evidence provided by correlations between population density and environmental humidity, like the one described in the introduction to this section; or will have to be confined to a consideration of stages in the life history where the insect has no opportunity of replenishing its water reserves, as for instance with pupae or other immobile developmental stages. In these it is generally found that high saturation deficits are associated with high mortalities, with different species showing marked differences in the level of resistance to desiccation. Coupled with measurements of the conditions to which developmental stages are subjected in the natural environment, such results may provide indications whether or not environmental humidity could be expected to have a marked effect on the rate of death among immature stages of the species concerned. Unless the analysis can be extended to active stages of the life history, however, there is little hope of assessing the magnitude of the over-all effect of humidity on the population in quantitative terms.

3. Effects of Humidity on the Rate of Birth

Humidity, like temperature, may affect the rate of birth in insect populations by influencing the rate at which offspring are produced, or the rate at which they complete their development. Oviposition rates have been shown to be sensitive to humidity in a number of insects, the precise relation between humidity and egg production differing from species to species (see Fig. 16.4(a)).

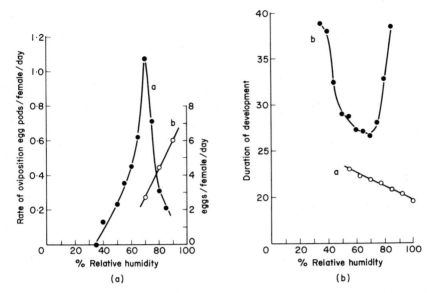

Fig. 16.4. The effect of humidity on the birth-rate of insects. (a) The effect of relative humidity on the rate of oviposition in *Locusta* (curve a, left ordinate) and in *Cryptolestes* (curve b, right ordinate) (Bursell, 1964 from Hamilton and from Ashby). (b) The effect of relative humidity on the duration of embryonic development in *Lucilia* (curve a, ordinate values in hours) and of nymphal development in *Locusta* (curve b, ordinate values in days) (Bursell, 1964 from Evans and from Hamilton).

Dry conditions are generally inimical to oviposition, but while some species are adversely affected by high humidities as well, in others the rate of oviposition increases progressively to reach a maximum in near-saturated atmospheres. The mechanisms of these effects have not yet been elucidated, but it is clear that they cannot be correlated in any simple way with the water content of egg-laying females.

Humidity also has a marked effect on the rate of development in many insects. Examples are shown in Fig. 16.4(b), where the duration of nymphal development in the locust is seen to be minimal at a relative humidity of 70%, increasing steeply both at higher and lower humidities. In the blowfly, the duration of embryonic development is seen to decrease progressively as the humidity is raised from 60% to 80% relative humidity.

The variety of effects on birth-rate illustrated in Fig. 16.4 precludes a general statement of the effect of humidity on the rate at which adults are recruited to a population of insects. Each species seems to show its own peculiarities in one way or another, and each must be investigated in its own right. All that can be said is that the rate of birth will, in many if not in most species of insect, be affected by humidity.

4. Humidities of Insect Habitats

Most terrestrial environments are markedly heterogeneous in respect of evaporating power, and they can usually be regarded as a mosaic of regions of high and low relative humidity. Evaporating power would be high in regions of open and relatively bare ground, for example, with high temperatures and correspondingly high saturation deficits associated with insolation, and with evaporative losses facilitated by unimpeded movements of air. In clumps of dense vegetation, on the other hand, air movement would be reduced, shade temperatures would be low and evaporation from plant and soil surfaces could produce microhabitats of negligible evaporating power. It has already been mentioned that insects are well equipped to take advantage of the variety of conditions offered by most terrestrial environments. Well-developed reactions to humidity have been demonstrated in a large number of species, and in many, the tactic reactions which lead to their aggregation in suitable parts of the environment are reinforced by kinetic mechanisms, which tend to confine the insects to suitable microenvironments once they have been located. With species that are relatively poorly water-proofed, and which would tend to be seriously threatened by prolonged exposure to desiccating conditions, a strong positive response to moist air is generally a constant feature of behaviour, as illustrated by curve (a) in Fig. 16.5. With species that are more resistant to desiccation, there is often an interesting reversal in the reaction to humidity. Individuals that have been given the opportunity of replenishing their water reserves show a marked preference for dry air (curves (b) and (c) of Fig. 16.5), but as their water reserves suffer depletion during exposure to their preferred humidity, so the dry preference becomes progressively weaker, and eventually it is replaced by a strong preference for humid air. Such an arrangement would ensure that the insects can take advantage, while their water reserves are plentiful, of the opportunities for feeding etc. offered by the general environment, taking refuge in more restricted regions of high relative humidity as necessity arises.

Such behavioural reactions of insects to environmental humidity will clearly complicate any attempt to assess the effect of humidity on population dynamics. One would need to obtain information about the proportion of time that the individuals of a species spend in microhabitats of humidity different from that of the general atmosphere, and such information is not readily available. Considerable technical difficulties are involved in monitoring continuously the humidity of appropriate microhabitats, and the daily pattern of activity in a population of insects living in their natural environment would be extremely difficult to assess. Reliance has usually to be placed on standard meteorological measurements in the hope that, though the results may not give an accurate estimate of the level of humidity to which the insects themselves are exposed, they may yet serve to provide a reasonable general indication of that level.

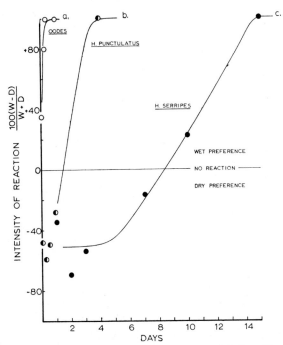

Fig. 16.5. Changes with time in the sign and intensity of the humidity reaction of three species of beetle, of which *Oodes* is the least and *Harpalus serripes* the most resistant to desiccation (Bursell, 1964 from Perttunen).

It is with humidity, in other words, as it is with temperature, that we are a long way from a thorough understanding of the relation between the physical feature of the environment and the population dynamics of the insect; and the difficulty is, that while the physiological effects of humidity, or temperature, can be reasonably well assessed, we are unable to specify precisely what are the humidities and temperatures to which the insect is subjected in the natural environment. The problem is thus, basically, an ecological one, and it is to be hoped that advances will be made in the field of ecological studies capable of overcoming the difficulty.

POSTSCRIPT

A review of the main features of insect physiology, such as has been attempted here, does as much to reveal our ignorance of the subject as it does to extol our knowledge. On every page there are gaps to show how far we still are from a thorough understanding of that particular metabolic system that is an insect, and how much further we need to go before we can hope to define what it is to be an insect. This is not to say that progress has been unimpressive. In the 20 years since Sir Vincent Wigglesworth, in his Croonian lecture, celebrated the insect as an experimental system for the study of physiology, a great deal has been done to establish the basic modes of operation of its parts. During this time advances in the field of insect biochemistry have been particularly spectacular, thanks to the development of a variety of biochemical microtechniques, and much has been learnt about the metabolic peculiarities that form the basis of an insect's life. In the field of somatic physiology, too, progress has been substantial, and a particularly striking feature of recent work is the consistent demonstration of regulative control in the operation of the different organ systems. The synthesis of digestive enzymes, for instance, has been shown to be controlled in relation to food intake; the loss of water from the respiratory system appears to be minutely regulated in relation to the state of water reserves; the resorption of water and of salts from the excretory system is carefully adjusted to meet the changing requirements of osmoregulation, and so on. The mechanisms by which the controls are exerted have in many cases been identified, but little has so far been discovered about the nature of the control systems themselves, and it is in this field that one may hope for substantial progress during the coming years. Indications have already been obtained that neuroendocrine elements are involved, but much work will need to be done to identify the specific components, and to elucidate the factors that control their activity.

Perhaps the greatest challenge that faces insect physiologists at the present day is that of extending their horizon beyond the laboratory, and of moving from a study of the individual insect exposed to controlled and usually constant conditions to the population of insects living in its natural habitat. Account would have to be taken of the spatial heterogeneity of the physical and biotic environment, of the diurnal and seasonal variations of its components, and of the capacity of insects to react appropriately, both at the physiological and at the behavioural level, to these differences in space and time. Formidable technical and conceptual difficulties would obviously have to be overcome, before such an extension of treatment could be achieved, and it is not surprising that progress to date has been negligible. It is possible, however, that an increasing awareness among insect physiologists and ecologists of the potentialities of computer technology for analysis of the complex interactions of the multitude of variables which would need to be considered may enable substantial advances to be made in this area. The practical implications of progress would be enormous in relation to the problems that face economic entomologists in the world today.

REFERENCES

Andrewartha, H. G. and Birch, L. C. (1954). "The Distribution and Abundance of Animals". University of Chicago Press, Chicago.
Baldwin, E. (1948). "Dynamic Aspects of Biochemistry". Cambridge University Press, Cambridge.
Barrass, R. (1961). *Behaviour* **18**, 288.
Beament, J. W. L. (1946). *Q. Jl. microsc. Sci.* **87**, 393.
Beament, J. W. L. (1949). *Bull. ent. Res.* **39**, 467.
Beament, J. W. L. (1959). *J. exp. Biol.* **36**, 391.
Beament, J. W. L. (1961). *Biol. Rev.* **36**, 281.
Berridge, M. J. (1965). *J. exp. Biol.* **43**, 535.
Berridge, M. J. (1968). *J. exp. Biol.* **48**, 159.
Bertram, D. S. and Bird, R. G. (1961). *Trans. R. Soc. trop. Med. Hyg.* **55**, 87.
Bouligand, Y. (1965). *C.r. hebd. Séanc. Acad. Sci. Paris* **261**, 3665-3668.
Brown, A. W. A. (1937). *J. exp. Biol.* **14**, 87.
Brunet, P. C. I. and Kent, P. W. (1955). *Proc. R. Soc.* B **144**, 259.
Buck, J. (1958). *Biol. Bull. mar. biol. Lab., Woods Hole* **114**, 118.
Bursell, E. (1957). *Proc. R. ent. Soc. Lond.* (A) **32**, 21.
Bursell, E. (1958). *Phil. Trans. R. Soc.* B **241**, 179.
Bursell, E. (1959). *Trans. R. ent. Soc. Lond.* **111**, 205.
Bursell, E. (1963). *J. Insect Physiol.* **9**, 439.
Bursell, E. (1964a). *J. Insect Physiol.* **10**, 993.
Bursell, E. (1964b). *In* "The Physiology of Insecta" (M. Rockstein, ed.), Vol. I. Academic Press, London and New York.
Burtt, E. T. and Catton, W. T. (1966). *Adv. Insect Physiol.* **3**, 1.
Chance, B. and Saktor, B. (1958). *Archs Biochem. Biophys.* **76**, 509.
Clever, U. (1963). *Devl. Biol.* **6**, 73.
Cockbain, A. J. (1961). *J. exp. Biol.* **38**, 163.
Davies, L. (1948). *J. exp. Biol.* **25**, 71.
Dethier, V. G. (1963). "The Physiology of Insect Senses". Methuen, London.
Duchâteau, G., Florkin, M. and Leclercq, J. (1953). *Archs int. Physiol.* **61**, 518.
Edney, E. B. (1966). *J. Insect Physiol.* **12**, 387.
Edney, E. B. and Barrass, R. (1962). *J. Insect Physiol.* **8**, 469.
Engelmann, F. (1957/58). *J. Insect Physiol.* **1**, 257.
Evans, D. R. and Dethier, V. G. (1957/58). *J. Insect Physiol.* **1**, 3.
Finlayson, L. H. and Lowenstein, O. (1958). *Proc. R. Soc.* B **148**, 433.
Finlayson, L. H. and Osborne, M. P. (1968). *J. Insect Physiol.* **14**, 1793.
Gilmour, D. (1961). "The Biochemistry of Insects". Academic Press, London and New York.
Glasgow, J. P. and Welch, J. R. (1962). *Bull. ent. Res.* **53**, 129.

Goldsmith, T. H. (1964). *In* "The Physiology of Insecta" (M. Rockstein, ed.), Vol. I. Academic Press, London and New York.

Gunn, D. L. and Pielou, D. P. (1940). *J. exp. Biol.* **17**, 307.

Harker, J. E. (1956). *J. exp. Biol.* **33**, 224.

Harrington, J. S. (1961). *Parasitology* **51**, 319.

Highnam, K. C. (1961). *Q. Jl. microsc. Sci.* **102**, 27.

Hodgson, E. S. (1964). *In* "The Physiology of Insecta" (M. Rockstein, ed.), Vol. I. Academic Press, London and New York.

Holdgate, M. W. and Seal, N. (1956). *J. exp. Biol.* **33**, 82.

Horridge, G. A. (1962). *Proc. Roy. Soc. B.* **157**, 38.

Horridge, G. A., Scholes, J. H., Shaw, S. and Tunstall, J. (1965). *In* "The Physiology of the Insect Central Nervous System" (J. E. Treherne and J. W. L. Beament, eds). Academic Press, London and New York.

Hoyle, G. (1960). *J. Insect Physiol.* **4**, 63.

Hoyle, G. (1965). *In* "The Physiology of the Insect Central Nervous System" (J. E. Treherne and J. W. L. Beament, eds). Academic Press, London and New York.

Hoyle, G. (1965). *In* "The Physiology of Insecta" (M. Rockstein, ed.), Vol. II. Academic Press, London and New York.

Huber, F. (1965). *In* "The Physiology of Insecta" (M. Rockstein, ed.), Vol. II. Academic Press, London and New York.

Hughes, G. M. (1965). *In* "The Physiology of the Insect Central Nervous System" (J. E. Treherne and J. W. L. Beament, eds). Academic Press, London and New York.

Imms, A. D. (1948). "A General Textbook of Entomology". Methuen, London.

Irreverre, F. and Terzian, L. A. (1959). *Science, N. Y.* **129**, 1358.

Ishizaki, H. (1965). *J. Insect Physiol.* **11**, 845.

Jackson, C. H. N. (1946). *Bull. ent. Res.* **37**, 291.

Jones, J. C. (1964). *In* "The Physiology of Insecta" (M. Rockstein, ed.), Vol. III. Academic Press, London and New York.

Karlson, P. and Sekeris, C. E. (1966). *Rec. Progr. Hormone Res.* **22**, 473.

Langley, P. A. (1966). *J. Insect Physiol.* **12**, 439.

Laughlin, P. (1957). *J. exp. Biol.* **34**, 226.

Lee, R. M. (1961). *J. Insect Physiol.* **7**, 37.

Levenbook, L. (1950). *Biochem. J.* **47**, 336.

Levy, R. I. and Schneiderman, H. A. (1958). *Nature, Lond.* **182**, 491.

Lewis, S. E. and Slater, E. C. (1954). *Biochem. J.* **58**, 207.

Lindauer, M. (1965). *In* "The Physiology of Insecta" (M. Rockstein, ed.), Vol. II. Academic Press, London and New York.

Locke, M. (1964). *In* "The Physiology of Insecta" (M. Rockstein, ed.), Vol. III. Academic Press, London and New York.

Loeb, J. and Northrup, J. H. (1917). *J. biol. Chem.* **32**, 103.

Lovelocke, J. E. (1953). *Biochim. biophys. Acta* **10**, 414.

Loveridge, J. P. (1968). *J. exp. Biol.* **49**, 15.

Lowenstein, O. and Finlayson, L. H. (1960). *Comp. Biochem. Physiol.* **1**, 56.

Markl, H. and Lindauer, M. (1965). *In* "The Physiology of Insecta" (M. Rockstein, ed.), Vol. III. Academic Press, London and New York.

McAllan, J. W. and Chefurka, W. (1961). *Comp. Biochem. Physiol.* **3**, 1.

Mead Briggs, A. R. (1956). *J. exp. Biol.* **33**, 737.

Miller, P. L. (1960). *J. exp. Biol.* **37**, 224, 237, 264.

Miller, P. L. (1964a). *In* "The Physiology of Insecta" (M. Rockstein, ed.), Vol. III. Academic Press, London and New York.

Miller, P. L. (1964b). *J. exp. Biol.* **41**, 331, 345.

Miller, P. L. (1965). *In* "The Physiology of the Insect Central Nervous System" (J. E. Treherne and J. W. L. Beament, eds). Academic Press, London and New York.

Mittelstaedt, H. (1962). *A. Rev. Ent.* **7**, 177.

Neville, A. C., Thomas, M. G. and Zelazny, B. (1969). *Tissue and Cell* **1**, 183-200.

Phelps, R. J. and Burrows, P. M. (1969). *Entomologia exp. appl.* **12**, 23.

Phillips, J. E. (1964a). *J. exp. Biol.* **41**, 15.

Phillips, J. E. (1964b). *J. exp. Biol.* **41**, 39.

Pringle, J. W. L. (1938). *J. exp. Biol.* **15**, 101, 114.

Pringle, J. W. L. (1965). *In* "The Physiology of Insecta" (M. Rockstein, ed.), Vol. II. Academic Press, London and New York.

Rajagopal. P. K. and Bursell, E. (1966). *J. Insect Physiol.* **12**, 287.

Ramsay, J. A. (1953). *J. exp. Biol.* **30**, 697.

Ramsay, J. A. (1954). *J. exp. Biol.* **31**, 104.

Ramsay, J. A. (1955). *J. exp. Biol.* **32**, 200.

Rowell, C. H. F. (1965). *In* "The Physiology of the Insect Central Nervous System" (J. E. Treherne and J. W. L. Beament, eds). Academic Press, London and New York.

Roeder, K. D. (1963). "Nerve Cells and Insect Behaviour". Harvard University Press, Cambridge, Mass., U.S.A.

Röller, H., Dahn, K. H., Sweeley, C. C. and Trost, B. M. (1967). *Angew. Chem.* **6**, 179.

Ruck, P. (1958). *J. Insect Physiol.* **2**, 189.

Schneider, D. (1962). *J. Insect Physiol.* **8**, 15.

Schneiderman, H. A. (1960). *Biol. Bull. mar. biol. Lab., Woods Hole* **119**, 494.

Schneiderman, H. A. and Williams, C. M. (1953). *Biol. Bull. mar. biol. Lab., Woods Hole* **105**, 320.

Schwartzkoppf, J. (1964). *In* "The Physiology of Insecta" (M. Rockstein, ed.), Vol. I. Academic Press, London and New York.

Smith, D. S. and Treherne, J. E. (1963). *Adv. Insect Physiol.* **1**, 401.

Stobbart, R. H. and Shaw, J. (1964). *In* "The Physiology of Insecta" (M. Rockstein, ed.), Vol. III. Academic Press, London and New York.

Sutcliffe, D. W. (1963). *Comp. Biochem. Physiol.* **9**, 121.

Thomsen, E. (1954). *J. exp. Biol.* **31**, 322.

Tinbergen, N., Meeuse, B. J. D., Boerema, L. K. and Karossieau, W. W. (1942). *Z. Tierpsychol.* **5**, 182.

Treherne, J. E. (1958). *J. exp. Biol.* **35**, 297.

Treherne, J. E. (1959). *J. exp. Biol.* **36**, 533.

Treherne, J. E. (1965a). *J. exp. Biol.* **42**, 7.

Treherne, J. E. (1965b). *In* "The Physiology of the Insect Central Nervous System" (J. E. Treherne and J. W. L. Beament, eds). Academic Press, London and New York.

Vowles, D. M. (1954). *J. exp. Biol.* **31**, 341.

Wallis, D. I. (1962). *Anim. Behav.* **10**, 105.

Weis Fogh, T. (1961). *In* "The Cell and the Organism" (J. A. Ramsay and V. B. Wigglesworth, eds). Cambridge University Press, Cambridge.

Wellington, W. G. (1949). *Scient. Agric.* **29**, 201.
Wigglesworth, V. B. (1931). *J. exp. Biol.* **8**, 411.
Wigglesworth, V. B. (1945). *J. exp. Biol.* **21**, 97.
Wigglesworth, V. B. (1952). *J. exp. Biol.* **29**, 561.
Wigglesworth, V. B. (1958). *Q. Jl. microsc. Sci.* **99**, 441.
Wigglesworth, V. B. (1959). "The Control of Growth and Form". Cornell University Press, Ithaca, New York.
Wigglesworth, V. B. (1960). *J. exp. Biol.* **37**, 500.
Wigglesworth, V. B. (1965). "The Principles of Insect Physiology". Methuen, London.
Wilde, J. de, (1964). *In* "The Physiology of Insecta" (M. Rockstein, ed.), Vol. I. Academic Press, London and New York.
Wilson, D. M. (1965). *In* "The Physiology of the Insect Central Nervous System" (J. E. Treherne and J. W. L. Beament, eds). Academic Press, London and New York.
Zebe, E., Delbrück, A. and Bücher, T. (1959). *Biochem. Z.* **331**, 254.

APPENDIX

INDEX TO INSECTS

AUTHOR INDEX

Numbers followed by an asterisk refer to the page on which the reference is listed

Wigglesworth, V. B., 50, 57, 77, 84, 114, 130, 187, 191, 194, 196, 202, 207, 212, 217, 247, 258*, 259*
Wilde, J., de, 184, 187, 191, 218, 259*
Williams, C. M., 92, 258*

Wilson, D. M., 179, 259*

Z

Zebe, E., 8, 259*
Zelazny, B., 24, 258*

SUBJECT INDEX

A

Absorption; alimentary, 38-40
  facilitation of, 39
  of fats, 40
  of glucose, 38-40
  of serine, 38-40
    rectal, 67-71
  of solutes, 70
  of water, 69
Acetate, 11
Acetylcholine, 52, 108
Acetyl CoA, 5, 11
Accessory glands, 184, 189, 191
Acclimation, 237
Adaptation, compound eye, 122, 124
Adenosine diphosphate (ADP), 3, 9
  in respiratory control, 10
Adenosine triphosphate (ATP), 3, 9, 10
Adipohaemocytes, 49
Adrenalin, 52
Air sacs, 84
Alanine, 5, 19, 42
Alary muscles, 50
Alimentary canal, 31-34
Allantoic acid, 21
  excretion of, 78-79
Allantoicase, 21
Allantoin, 21
  excretion of, 78-79
  in haemolymph, 47
Allantoinase, 21, 78
Ametabola, 199
  development of, 200
Amino acids
  excretion of, 79-80
  in haemolymph, 17, 45-46
  in nutrition, 42
  oxidative deamination of, 16-19
Ammonia
  excretion of, 79
  metabolism of, 20-23

Amylase, 35
Aorta, 50
Apposition eyes, 122, 124
Arginase, 79
Arginine, 42
  excretion of, 80
Arrest, of development, 204-207
Aspartic acid, 5, 19, 42, 46
Atropine, 52
Auditory organs, 133, 143-144
Axon, 107-108, 113, 115

B

Bacteria, 41
Basement membrane, formation of, 195
Behaviour, 157-180
Bicarbonate
  in haemolymph, 87
  in respiratory regulation, 87, 92
Biotin, 43
Birth rate
  effect of humidity on, 250-251
  effect of temperature on, 234-235
Body temperature, 231-234
Brain hormone, 211
Buccal cavity, 33

C

Caecae, gastric, 33, 39
Calcium, in haemolymph, 45, 62
Campaniform sensillae, 130
Carbohydrases, 34-35
Carbohydrate
  absorption of, 38-40
  digestion of, 34-35
  metabolism of, 12-16
Carbon dioxide
  effect of on spiracles, 89-90, 148
  effect of on ventilation, 94